COMPARATIVE ANATOMY OF DOMESTIC ANIMALS

A Guide

COMPARATIVE ANATOMY OF DOMESTIC ANIMALS

A Guide

BONNIE BEAVER

The Iowa State University Press / Ames

Bonnie Beaver is associate professor, veterinary anatomy, College of Veterinary Medicine, Texas A&M University.

Printed by
The Iowa State University Press
Ames, Iowa 50010

First edition, 1980

International Standard Book Number: 0-8138-1545-2

CONTENTS

ILLUSTRATIONS

PREFACE

Comparative Anatomy of Domestic Animals: A Guide was written to serve as an aid in teaching veterinary gross and clinical anatomy and to serve as a reference of anatomical structures of the domestic animals. The terminology is that of the *Nomina Anatomica Veterinaria*, 1972, with popular synonyms and species differences in parentheses. The guide is not intended to include all anatomical structures but rather to contain structures of clinical significance and to compare those with other species. Clinical relevance of the structures and areas is left to the instructor.

The author wishes to express thanks to Dr. R.F. Sis and to the faculty and staff of the Department of Veterinary Anatomy, Texas A&M University for their help in this venture. Special appreciation and thanks go to Sharon Ashby and Dr. Ed Smallwood for the illustrations and to Elizabeth Pannill, Pat Stroud, Dr. Mike Tatum, and Olgui Urso for the manuscript.

COMPARATIVE ANATOMY OF DOMESTIC ANIMALS

A Guide

1

THORACIC LIMB

OSTEOLOGY

Scapula
- Spine
  - Acromion (-BCFp)
    - Hamate process (---F-)
    - Suprahamate process (---F-)
  - Tuber spinae (E---P)
- Supraspinous fossa
- Infraspinous fossa
- Neck
- Supraglenoid tubercle
  - Coracoid process
- Infraglenoid tubercle (--CF-)
- Subscapular fossa
- Serrated face
- Glenoid cavity
- Scapular cartilage

Humeral (shoulder) joint

Clavicle (--cF-)

Humerus
- Head
- Neck
- Major (greater) tubercle
- Minor (lesser) tubercle
- Intertubercular groove
  - Intermediate tubercle (E----)
- Crest of the major tubercle
  - Deltoid tuberosity
  - Teres minor tuberosity
  - Tricipital line
- Teres major tuberosity
- Brachial groove

Structures mentioned in the text are present in the equine (E), bovine (B), canine (C) and feline (F) species unless a minus sign (-) indicates otherwise. Structures present but of minor significance are indicated by the lowercase letters (e), (b), (c), and (f). Porcine (P) structures are also include in the osteology, myology, oral, digestive, and urogenital sections.

Humerus (cont.)
- Medial epicondyle
  - Supracondylar foramen (---F-)
- Lateral epicondyle
  - Lateral epicondyloid crest
- Condyle
  - Supratrochlear foramen (--C--)
  - Trochlea
  - Capitulum
  - Radial fossa
  - Coronoid fossa (---F-)
  - Olecranon fossa
- Nutrient foramen

Cubital (elbow) joint

Radius
- Fused with ulna (EB---)
- Head
  - Fovea capitis
  - Articular circumference
- Radial tuberosity
- Ulnar notch (--CFP)
- Trochlea
  - Styloid process (medial)
- Interosseous space (EB---)
- Groove for extensor carpi radialis tendon
- Groove for common digital extensor tendon
- Groove for abductor pollicis longus tendon

Ulna
- Fused with radius (EB---)
- Olecranon
  - Tuber olecrani
  - Anconeal process
- Trochlear notch
- Coronoid processes
  - Radial notch (--CFP)
- Styloid process (lateral)
- Nutrient foramen

Antebrachiocarpal joint

Carpal bones
- Proximal row
  - Radial (EB--P)
  - Intermediate (EB--P)
  - Intermedioradial (--CF-)
  - Ulnar
  - Accessory

- Middle carpal joint

- Distal row
  - First (E-CFP)
  - Second
    - Fused with third (-B---)
  - Third
    - Fused with second (-B---)
  - Fourth
  - Fifth (e----)
  - Sesamoid of abductor pollicis longus m. (--CF-)

Carpometacarpal joint

Metacarpal bones
- First (--CF-)
- Second (E-CFP)
- Third
  - Fused with fourth (-B---)
  - Dorsal longitudinal groove (-B---)
  - Palmar longitudinal groove (-B---)
  - Intertrochlear notch (-B---)
- Fourth
  - Fused with third (-B---)
- Fifth (-BCFP)

Metacarpophalangeal joints

Proximal (palmar) sesamoid bones
- Articular surface
- Flexor surface
- Interosseous surface

Dorsal sesamoid bones (--CF-)

Phalanges
- Proximal phalanx
- Proximal interphalangeal joint
- Middle phalanx
- Distal interphalangeal joint
- Distal (palmar) sesamoid bone (EB--P)
- Distal phalanx
  - Axial foramen (-B---)
  - Extensor process (EB--P)
  - Marginal notch (crena) (E----)
  - Medial and lateral palmar processes (EB--P)
  - Parietal grooves (E----)
  - Semilunar line (E----)
  - Solar foramen to canal of the sole (E----)
  - Base (--CF-) (Fig. 1)
  - Ungual crest (--CF-) (Fig. 1)
  - Ungual process (--CF-) (Fig. 1)
- Phalanges of the dewclaws (-B--P)

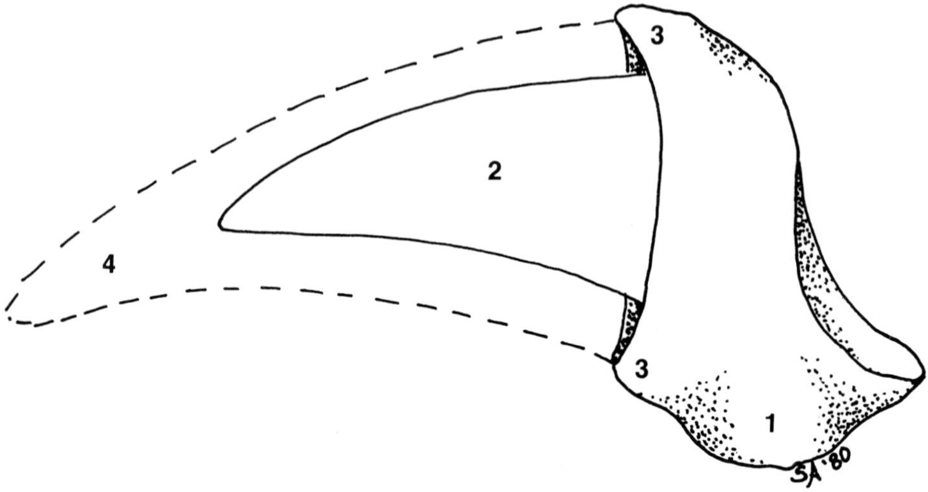

Figure 1. Distal phalanx of the dog and cat (lateral aspect)

1. Base of distal phalanx
2. Ungual process of distal phalanx
3. Ungual crest of distal phalanx
4. Claw

MYOLOGY

Extrinsic muscles

- Trapezius m.
  - Origin
    - Dorsal neck (cervical portion)
    - Dorsal thorax (thoracic portion)
  - Termination
    - Spine of scapula
  - Function
    - Elevate limb
  - Innervation
    - Accessory n.
  - Blood supply
    - Dorsal scapular a. (EB--P)
    - Deep cervical a. (EB---)
    - Superficial cervical a. (--CF-)

- Rhomboideus m.
  - Origin
    - Dorsal neck (cervical portion)
    - Dorsal thorax (thoracic portion)
    - Nuchal crest of skull (capital portion) (--CFP)
  - Termination
    - Dorsal scapula and cartilage
  - Function
    - Elevate limb
    - Forms Brahman bull hump (-B---)
  - Innervation
    - Cervical and thoracic nn.
  - Blood supply
    - Dorsal scapular a. (EB--P)
    - Deep cervical a. (EB---)
    - Superficial cervical a. (--CF-)

- Omotransversarius m.
  - Origin
    - Wings of atlas
    - Cervical vertebrae (E----)
  - Termination
    - Spine of scapula (ebCFp)
  - Function
    - Draw limb forward
    - Depress head and neck
  - Innervation
    - Accessory n.
  - Blood supply
    - Superficial cervical a.

- Latissimus dorsi m.
  - Origin
    - Dorsal thoracolumbar area

Extrinsic muscles, latissimus dorsi m. (cont.)

- Termination
    - Teres major tuberosity (EBCF-)
    - Minor tubercle (----P)
- Function
    - Elevate limb
    - Draw limb caudally
    - Flex humeral joint
- Innervation
    - Thoracodorsal n.
- Blood supply
    - Thoracodorsal a.
    - Thoracic and lumbar aa.

Brachiocephalicus m.

- Proximal attachment (cleidocephalicus m.)
    - Mastoid process of skull (cleidomastoideus m.)
    - Dorsal cervical area (cleidocervicalis m.) (--CF-)
    - Occipital bone (cleidooccipitalis m.) (-B--P)
- Distal attachment (cleidobrachialis m.)
    - Distocranial humerus (EBC-P)
    - Proximal ulna (---F-)
- Function
    - Draw limb forward
    - Depress head and neck
    - Covers cranial humeral joint (E-CFP)
- Innervation
    - Accessory n.
    - Axillary n.
    - Cervical nn.
- Blood supply
    - Superficial cervical a.
    - Deep cervical a. (-B---)

Superficial pectoral m.

- Descending (cranial superficial) pectoral m.
    - Origin
        - Sternum
    - Termination
        - Crest of the major tubercle (EBC--)
        - Proximal ulna (---F-)
    - Function
        - Adduct limb
        - Forms bulge of chest (E----)
    - Innervation
        - Cranial pectoral nn.
    - Blood supply
        - External thoracic a.
- Transverse (caudal superficial) pectoral m.
    - Origin
        - Sternum

Termination
- Distocranial humerus
- Proximal ulna (---F-)

Function
- Adduct limb
- Forms "V" of chest (E----)

Blood supply
- External thoracic a.

Subclavius (cranial deep pectoral) m. (Eb--P)

Origin
- Sternum

Termination
- Lateral scapular fascia

Function
- Adduct limb

Innervation
- Caudal pectoral nn.

Blood supply
- Superficial cervical a.
- External thoracic a.

Deep (ascending) (caudal deep) pectoral m.

Origin
- Sternum

Termination
- Major tubercle (EBC-P)
- Minor tubercle (EBC-P)
- Crest of the major tubercle (---F-)

Function
- Adduct limb
- Retract limb

Innervation
- Caudal pectoral nn.

Blood supply
- External thoracic a.
- Lateral thoracic a.
- Thoracic aa.

Serratus ventralis m.

Origin
- Ventral rib bones (thoracic portion)
- Cervical transverse processes (cervical portion)

Termination
- Serrated face

Function
- Support trunk between the limbs
- Elevate trunk between the limbs

Innervation
- Long thoracic n.
- Cervical nn.

Extrinsic muscles, serratus ventralis m. (cont.)
- Blood supply
  - Cervical and thoracic aa.
  - Dorsal scapular a. (-B--P)
  - Deep cervical a. (EB---)

Intrinsic muscles
- Deltoideus m.
  - Origin
    - Spine of scapula (scapular portion)
    - Acromion (acromial portion) (-BCF-)
  - Termination
    - Deltoid tuberosity
  - Function
    - Flex humeral joint
  - Innervation
    - Axillary n.
  - Blood supply
    - Subscapular a.

- Supraspinatus m.
  - Origin
    - Supraspinous fossa
  - Termination
    - Major tubercle
    - Minor tubercle (----P)
  - Function
    - Extend humeral joint
    - Abduct lower limb
  - Innervation
    - Suprascapular n.
  - Blood supply
    - Superifcial cervical a. (EBCF-)
      - Suprascapular a.
    - Suprascapular a. (----P)
  - Bursa
    - As it crosses the major tubercle (----P)

- Infraspinatus m.
  - Origin
    - Infraspinous fossa
  - Termination
    - Major tubercle
  - Function
    - Serve as lateral collateral lig.
    - Abduct lower limb
    - Flex humeral joint
  - Innervation
    - Suprascapular n.
  - Blood supply
    - Subscapular a.
    - Suprascapular a. (----P)

Bursa
As it crosses the major tubercle (EBC?P)

Teres minor m.
Origin
Infraglenoid tubercle
Termination
Teres minor tuberosity
Function
Flex humeral joint
Innervation
Axillary n.
Blood supply
Caudal circumflex humeral a.
Superficial cervical a. (--CF-)

Subscapularis m.
Origin
Subscapular fossa
Termination
Minor tubercle
Function
Serve as medial collateral lig.
Adduct lower limb
Innervation
Subscapular n.
Blood supply
Subscapular a.

Teres major m.
Origin
Caudal border of scapula
Termination
Teres major tuberosity
Function
Flex humeral joint
Innervation
Axillary n.
Blood supply
Subscapular a.

Coracobrachialis m.
Origin
Coracoid process of scapula
Termination
Proximomedial humerus
Innervation
Musculocutaneous n.
Blood supply
Cranial circumflex humeral a.
Caudal circumflex humeral a. (-BCF-)

Intrinsic muscles (cont.)

- Biceps brachii m.
  - Origin
    - Supraglenoid tubercle
  - Termination
    - Radial tuberosity
    - Proximal ulna (--C-P)
  - Function
    - Extend humeral joint
    - Flex cubital joint
    - Serve as cranial lig. of the shoulder
  - Innervation
    - Musculocutaneous n.
  - Blood supply
    - Bicipital a.
    - Cranial circumflex humeral a. (-BCFP)
    - Transverse cubital a. (E---P)
  - Bursa
    - Intratubercular bursa (EB---)
  - Miscellaneous
    - Lacertus fibrosus (E----)

- Brachialis m.
  - Origin
    - Proximolateral humerus
  - Termination
    - Radial tuberosity (EBC-P)
    - Proximal ulna (--CFP)
  - Function
    - Flex cubital joint
  - Innervation
    - Musculocutaneous n.
  - Blood supply
    - Caudal circumflex humeral a.
  - Bursa
    - Between ulnar tendon and ulna (----P)

- Tensor fasciae antebrachii m.
  - Origin
    - Latissimus dorsi m.
  - Termination
    - Medial fascia of forearm
    - Olecranon
  - Innervation
    - Radial n.

- Triceps brachii m.
  - Long head
    - Origin
      - Caudal border of scapula
    - Function
      - Flex humeral joint

Extend cubital joint
Innervation
Radial n.
Blood supply
Caudal circumflex humeral a.
Deep brachial a.
Bursa
Superficial bursa between skin and tendon
Deep bursa between tendon and olecranon
Lateral head
Origin
Tricipital line
Termination
Olecranon
Function
Extend cubital joint
Innervation
Radial n.
Blood supply
Caudal circumflex humeral a.
Deep brachial a.
Bursa
Between tendon and olecranon (----P)
Medial head
Origin
Proximomedial humerus
Termination
Olecranon
Function
Extend cubital joint
Covers structures of supracondylar foramen (---F-)
Bursa
Between tendon and olecranon (----P)
Accessory head (--CF-)
Origin
Neck of humerus
Termination
Olecranon
Function
Extend cubital joint
Innervation
Radial n.
Blood supply
Caudal circumflex humeral a.
Deep brachial a.

Anconeus m.
Origin
Lateral epicondyle and epicondylar crest
Termination
Lateral surface of olecranon

Intrinsic muscles, anconeus m. (cont.)
- Innervation
  - Radial n.
- Miscellaneous
  - Double (----P)

Brachioradialis m. (--CF-)
- Covers cephalic v.
- Innervation
  - Radial n.

Extensor carpi radialis m.
- Origin
  - Lateral epicondyle and epicondylar crest
- Termination
  - Proximodorsal metacarpal(s)
- Function
  - Flex cubital joint
  - Extend carpus
- Innervation
  - Radial n.
- Blood supply
  - Transverse cubital a.

Common digital extensor m.
- Origin
  - Lateral epicondyle
  - Ulna (-B---)
- Termination
  - Distal phalanx
  - Middle phalanx of third digit (EB---)
  - Proximal phalanx of third digit (EB---)
- Function
  - Extend carpus
  - Extend digits
- Innervation
  - Radial n.
- Blood supply
  - Cranial interosseous a.
  - Transverse cubital a.
  - Caudal interosseous a. (-BCFP)
- Sesamoid bones
  - Dorsal sesamoid bone in tendon (--CF-)

Lateral digital extensor m.
- Origin
  - Proximolateral radius (EB---)
  - Proximolateral ulna (-B---)
  - Lateral collateral lig. of cubital joint (-B---)
  - Lateral epicondyle of humerus (--CFP)
- Termination
  - Proximal phalanx (E----)

Middle phalanx of fourth digit (-B---)
Distal phalanx of third digit (--CF-)
Distal phalanx of fourth digit (-BCFP)
Distal phalanx of fifth digit (--CFP)

Function
Extend digits

Innervation
Radial n.

Blood supply
Cranial interosseous a.
Caudal interosseous a. (-BCF-)
Transverse cubital a. (EB---)

Abductor pollicis longus (abductor digiti I longus) m.

Origin
Distolateral radius (EB--P)
Craniolateral ulna (-BCFP)

Termination
Proximal metacarpals

Function
Extend carpus
Abduct first digit (--CF-)

Innervation
Radial n.

Blood supply
Cranial interosseous a. (EB--P)
Caudal interosseous a. (--CF-)

Sesamoid
Sesamoid in the tendon at medial carpus (--CF-)

Supinator m. (--CFp)

Origin
Lateral epicondyle (--C-p)
Lateral collateral lig. of cubital joint (---F-)

Termination
Proximocranial radius

Function
Supinate manus

Innervation
Radial n.

Blood supply
Transverse cubital a.

Extensor carpi ulnaris (ulnaris lateralis) m.

Origin
Lateral epicondyle

Termination
Proximolateral metacarpus
Accessory carpal bone (EBC--)

Function
Flex carpus

Intrinsic muscles, extensor carpi ulnaris m. (cont.)

Innervation
- Radial n.

Blood supply
- Cranial interosseous a.
- Caudal interosseous a. (----P)

Flexor carpi ulnaris m.

Origin
- Medial epicondyle (humeral head)
- Olecranon (ulnar head) (EBCFp)

Termination
- Accessory carpal bone

Function
- Flex carpus
- Extend cubital joint

Innervation
- Ulnar n.

Blood supply
- Collateral ulnar a.
- Ulnar a. (--CF-)
- Cranial interosseous a. (-B---)
- Median a. (E----)

Flexor carpi radialis m.

Origin
- Medial epicondyle

Termination
- Proximopalmar metacarpal(s)

Function
- Flex carpus

Innervation
- Median n.

Blood supply
- Collateral ulnar a. (-BCFP)
- Median a. (E----)

Superficial digital flexor m.

Origin
- Medial epicondyle

Termination
- Proximal phalanx (E----)
- Middle phalanges

Function
- Flex carpus
- Flex digits

Innervation
- Median n.

Blood supply
- Collateral ulnar a.
- Median a.
- Caudal interosseous a. (EB--P)

- Miscellaneous
    - Forms manica flexoria for deep digital flexor tendon
- Equine modifications
    - Accessory lig. (radial head) (superior check lig.) (Fig. 6)
        - Connects caudal radius to superficial digital flexor tendon
- Bovine modifications
    - Superficial portion
        - Superficial to flexor retinaculum
    - Deep portion
        - Within carpal canal
        - Receives branch (interflexorius m.) from deep digital flexor m.
        - Rejoins superficial portion at midmetacarpus
    - Interosseous m. sends branch to each side of tendons
- Feline modifications
    - Tendon in metacarpal region is fleshy
- Porcine modifications
    - Superficial portion
        - Superficial to flexor retinaculum
        - Attaches to fourth digit
    - Deep portion
        - Within carpal canal
        - Receives branch from deep digital flexor m.
        - Attaches to third digit

Deep digital flexor m.

- Origin
    - Medial epicondyle (humeral head)
    - Olecranon (ulnar head) (EB--P)
    - Caudal ulna (ulnar head) (--CF-)
    - Medial radius (radial head)
- Termination
    - Distal phalanges
- Function
    - Flex carpus
    - Flex digits
- Innervation
    - Humeral and ulnar heads
        - Median n.
        - Ulnar n.
    - Radial head
        - Median n.
- Blood supply
    - Collateral ulnar a.
    - Caudal interosseous a. (-BCF-)
- Equine Modifications
    - Accessory lig. (inferior check lig.) (Fig. 6)
        - Connects carpal annular lig. to deep digital flexor tendon
    - Navicular bursa

Intrinsic muscles, deep digital flexor m. (cont.)
- Bovine modifications
  - Humeral head sends branch (interflexorius m.) to superficial digital flexor m.
- Feline modifications
  - Interflexorius m.
    - Origin
      - Humeral head of deep digital flexor m.
    - Termination
      - Middle phalanx
    - Function
      - Flex carpus
      - Flex digits
    - Innervation
      - Median n.
    - Present in dogs but not well developed
- Porcine modifications
  - Humeral head receives fibers from superficial digital flexor m.
  - Sends branch to superficial digital flexor m.

Pronator teres m.
- Origin
  - Medial epicondyle
- Termination
  - Medial radius
- Function
  - Flex cubital joint
  - Pronate manus
- Innervation
  - Median n.
- Blood supply
  - Median a.

Pronator quadratus m. (--CF-)
- Attachments
  - Between radius and ulna
- Function
  - Pronate manus
- Innervation
  - Median n.
- Blood supply
  - Caudal interosseous a.

Lumbricales m. (--cFP)
- Origin
  - Tendon of deep digital flexor m.
- Termination
  - Proximal phalanx
  - Deep portion of superficial digital flexor m. (----P)
- Function
  - Flex digits

Intrinsic muscles (cont.)
- Interosseous mm.
  - Origin
    - Palmar carpal lig.
    - Proximal metacarpals
  - Termination
    - Proximal palmar sesamoids
    - Common digital extensor tendon
  - Function
    - Flex metacarpophalangeal joint
  - Innervation
    - Ulnar n.

LIGAMENTS

Dorsoscapular lig. (E---)
- Boney attachments
  - Dorsal spines of T_{3-5}
  - Serrated face of scapula
- Muscles arising from the ligament
  - Rhomboideus m.
  - Semispinalis capitis m.
  - Splenius m.
  - Serratus dorsalis cranialis m.
- Muscles sheathed by the lamellae
  - Spinalis et semispinalis m.
  - Longissimus dorsi m.
  - Iliocostalis m.
- Functions
  - Prevents proximal scapula from abducting
- Vessels in the area
  - Branches of costocervical a. & v.
- Supraspinous bursa
  - Protects nuchal lig. from T_{2-4} dorsal spines and dorsoscapular lig.

Transverse humeral lig. (--CF)
- Crosses intertubercular groove
- Holds biceps brachii tendon in groove

Medial and lateral collateral ligg. of the cubital joint
- Medial and lateral epicondyles of humerus
- Proximal medial and lateral radius and ulna

Medial and lateral collateral ligg. of the carpus

Short carpal ligg. (Figs. 2,3)
- Between individual carpal bones

Extensor retinaculum of the carpus

Flexor retinaculum of the carpus
- Connects accessory carpal bone to radial and fourth carpal bones

Palmar carpal lig.
- Fits closely to palmar side of carpal bones
- Gives origin to
  - Interosseus m. (suspensory lig.)
  - Accessory lig. to the deep digital flexor m. (E---)

Carpal canal
- Between flexor retinaculum and palmar carpal lig.
- Contains
  - Flexor carpi radialis tendon (E---)
  - Superficial digital flexor tendon
  - Deep digital flexor tendon
  - Radial a. & v.
  - Ulnar a. & v.
  - Caudal interosseous a. & v.
  - Ulnar n. (-BCF)
  - Median n. (-BCF)
  - Medial and lateral palmar (common) (digital) nn. (E---)

Medial and lateral collateral ligg. of the metacarpophalangeal joint (Figs. 2,3,4,7)
- Proximal attachment
  - Distal metacarpals
- Distal attachment
  - Proximal part of proximal phalanx
    - Superficial portion (E---)
  - Deep portion (E---)
    - Distal abaxial surface of proximal sesamoids (E---)

Interphalangeal collateral ligg. (medial and lateral or axial and abaxial) (Figs. 2,3,4,7)
- Locations
  - Proximal interphalangeal joints
  - Distal interphalangeal joints

Palmar ligg. of the proximal interphalangeal joint (EB--) (Figs. 5,8)
- Medial and lateral
  - Proximal attachment
    - Distal aspect of proximal phalanx
  - Distal attachment
    - Proximopalmar aspect of middle phalanx
- Central
  - Proximal attachment
    - Distopalmar aspect of proximal phalanx
  - Distal attachment
    - Proximopalmar aspect of middle phalanx
  - Paired ligg. (E---)

Proximal interdigital ligg. (-BCF) (Fig. 7)
    Location
        Between proximal phalanges
    Function
        Prevent undue divergence of phalanges

Collateral lig. of the distal sesamoid (Eb--) (Fig. 5)
    Proximal attachment
        Distal medial and lateral aspects of proximal phalanx
    Distal attachment
        Proximal border of distal sesamoid
    Function
        Forms a "U" shaped suspending attachment

Distal sesamoidean lig. (EB--) (Fig. 5,8)
    Proximal attachment
        Distal border of distal sesamoid
    Distal attachment
        Palmar aspect of distal phalanx

Distal interdigital ligg. (-B--) (Fig. 8)
    Proximal attachment
        Axial, proximopalmar aspect of middle phalanx
    Distal attachment
        Axial, proximopalmar aspect of distal phalanx of opposite digit
    Function
        Limit abaxial movement of palmar part of foot

Dorsal band (-B--) (Fig. 7)
    Proximal attachment
        Distal abaxial aspect of proximal phalanx
    Distal attachment
        Extensor process of distal phalanx

Dorsal lig. (--C-) (Figs. 2,3)
    Proximal attachment
        Proximal axial and abaxial aspects of middle phalanges
    Distal attachment
        Proximodorsal aspect of distal phalanges
    Function
        Oppose deep digital flexor m. to keep claw off ground

Collateral ligg. (Figs. 2,3,5,8)
    Proximal attachment
        Abaxial surface of each proximal sesamoid
    Distal attachments
        Distal metacarpal bones
        Proximal aspect of proximal phalanges

Strong elastic lig. (---F) (Fig. 3)
- Proximal attachment
  - Distolateral aspect of middle phalanges
- Distal attachment
  - Dorsal ungual crest of distal phalanges
- Function
  - Claw retraction
  - Distal phalanx is pulled to lateral side of middle phalanx

Suspensory apparatus of the metacarpophalangeal joint
- Interosseous m. (suspensory lig.) (EB--) (Figs. 6,9)
  - Proximal attachments
    - Palmar carpal lig.
    - Proximal metacarpus
  - Distal branches
    - Medial and lateral branch to superficial digital flexor tendon to help form manica flexoria (-B--)
    - Medial and lateral branches to common digital extensor tendon
    - Central branch (-B--)
      - Branch to axial side of proximal sesamoids
      - Axial branch to common digital extensor tendon
  - Function
    - Support metacarpophalangeal joint
    - Prevent hyperextension of digit

- Palmar lig. of the metacarpophalangeal joint (Figs. 2,5,8)
  - Embeds proximal (palmar) sesamoids

- Straight (superficial) sesamoidean lig. (E---) (Fig. 6)
  - Proximal attachment
    - Base of proximal sesamoid bones
    - Intersesamoidean lig.
  - Distal attachment
    - Proximal palmar aspect of middle phalanx

- Oblique (middle) sesamoidean lig. (EB--) (Figs. 6,9)
  - Proximal attachment
    - Base of proximal sesamoid bones
  - Distal attachment
    - Proximopalmar aspect of proximal phalanx (-B--)
    - Midpalmar aspect of proximal phalanx (E---)

- Cruciate (deep) sesamoidean lig. (Figs. 2,6,9)
  - Proximal attachment
    - Base of proximal (palmar) sesamoid bones
  - Distal attachment
    - Proximopalmar aspect of proximal phalanx on side opposite origin

- Short sesamoidean lig. (E---) (Fig. 4)
  - Proximal attachment
    - Base of proximal sesamoid bones

Distal attachment
Proximopalmar aspect of proximal phalanx on same side as origin

Palmar annular lig. of the metacarpophalangeal joint

Proximal digital annular lig.
Attaches to proximal phalanx
Encloses digital flexor tendon sheath

Distal digital annular lig.
Attaches to middle phalanx
Encloses digital flexor tendon sheath

Lig. of the ergot (E---)
Proximal attachment
Base of ergot
Distal attachment
Blends with distal digital annular lig.
Can be confused with palmar (proper) digital nn.

SYNOVIAL SHEATHS

Carpal synovial tendon sheaths (Fig. 10)
7 of 9 muscles crossing the carpus are usually associated with a tendon sheath
Extensor carpi radialis tendon sheath
Common digital extensor tendon sheath
Lateral digital extensor tendon sheath
Extensor carpi ulnaris tendon sheath (E---)
Flexor carpi ulnaris m. (no tendon sheath)
Flexor carpi radialis tendon sheath
Abductor pollicis longus tendon sheath
Superficial digital flexor tendon sheath
Shared with
Deep digital flexor m. (E---)
Deep digital flexor tendon sheath
Shared with
Superficial digital flexor m.(E---)
Proximal position
Usually begin just proximal to carpus
Distal position
Usually end just distal to carpus
Digital flexor tendon sheath ends at midmetacarpus (E---)

Digital synovial tendon sheath
Surrounds superficial and deep digital flexor tendons
Proximal position
Usually begin at distal 1/4 of metacarpus
Lowest point of access (E---)
Proximal edge of distal digital annular lig. as it crosses deep digital flexor m.

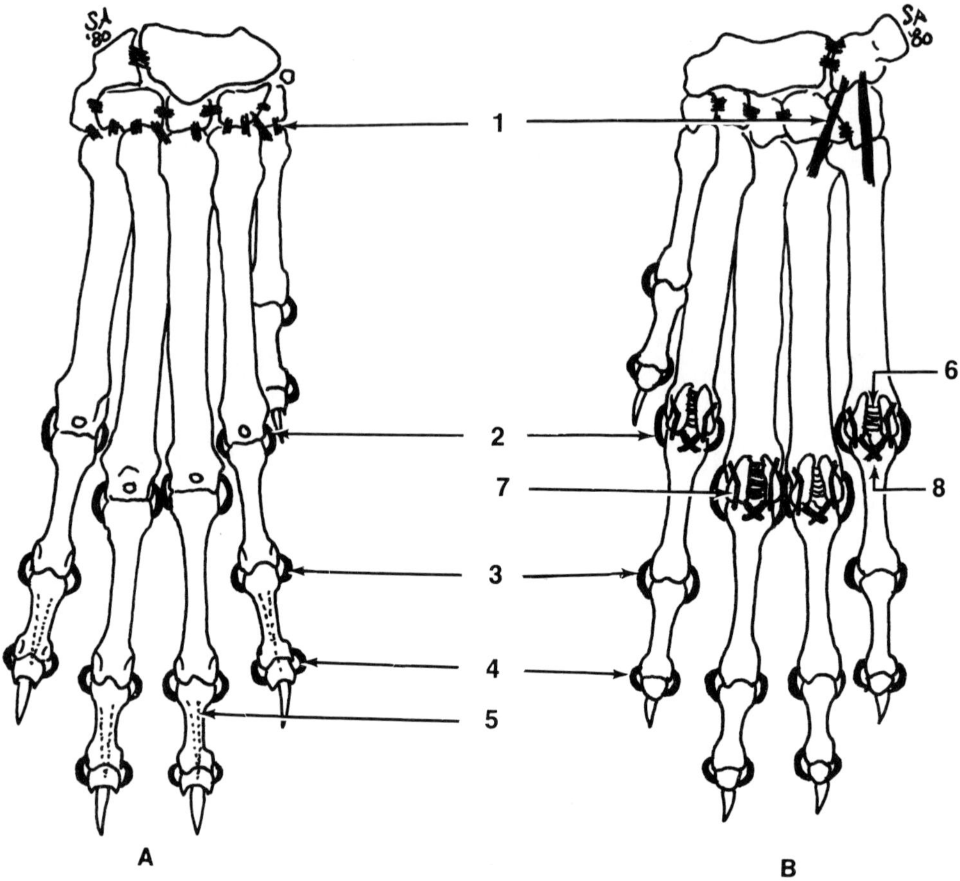

Figure 2. Ligaments of the canine and feline thoracic limb (A-dorsal aspect, B-palmar aspect)

1. Short carpal ligaments
2. Metacarpophalangeal collateral ligaments
3. Proximal interphalangeal collateral ligaments
4. Distal interphalangeal collateral ligaments
5. Dorsal ligaments
6. Palmar ligaments of metacarpophalangeal joint
7. Collateral sesamoidean ligaments
8. Cruciate sesamoidean ligaments

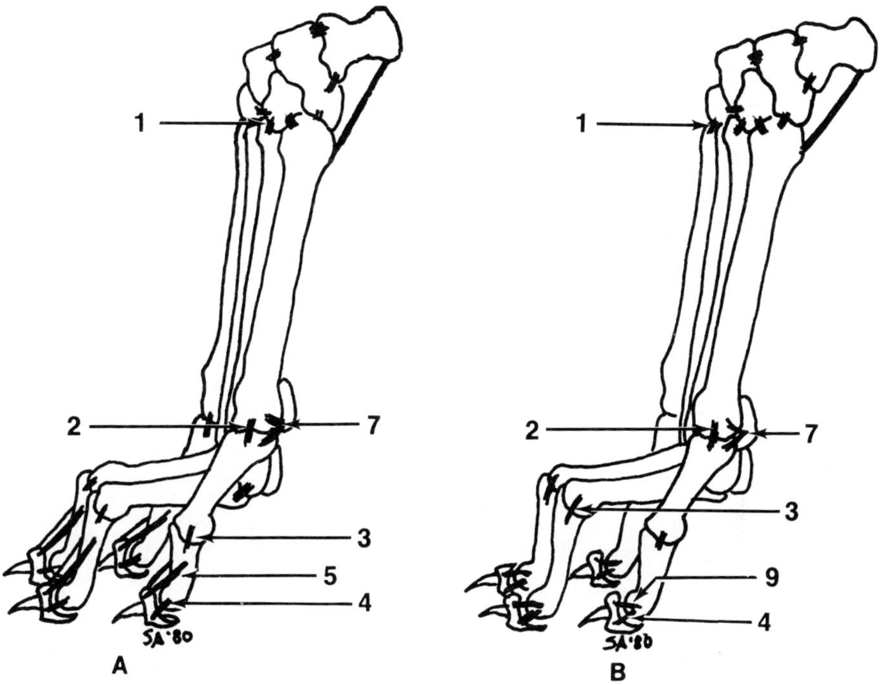

Figure 3. Ligaments of the canine and feline thoracic limb (lateral aspect, A-dog, B-cat)

1. Short carpal ligaments
2. Metacarpophalangeal collateral ligaments
3. Proximal interphalangeal collateral ligaments
4. Distal interphalangeal collateral ligaments
5. Dorsal ligaments
7. Collateral sesamoidean ligaments
9. Strong elastic ligaments

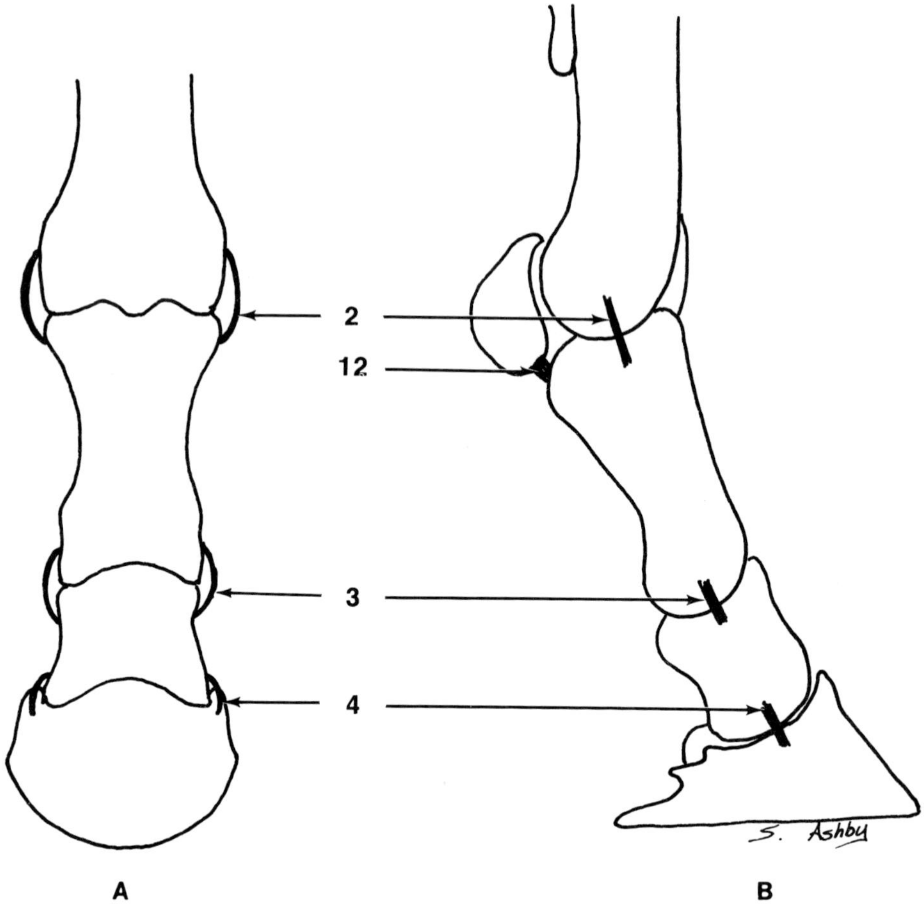

Figure 4. Ligaments of the equine thoracic limb (A-dorsal aspect, B-lateral aspect)

2. Metacarpophalangeal collateral ligaments
3. Proximal interphalangeal collateral ligaments
4. Distal interphalangeal collateral ligaments
12. Short sesamoidean ligaments

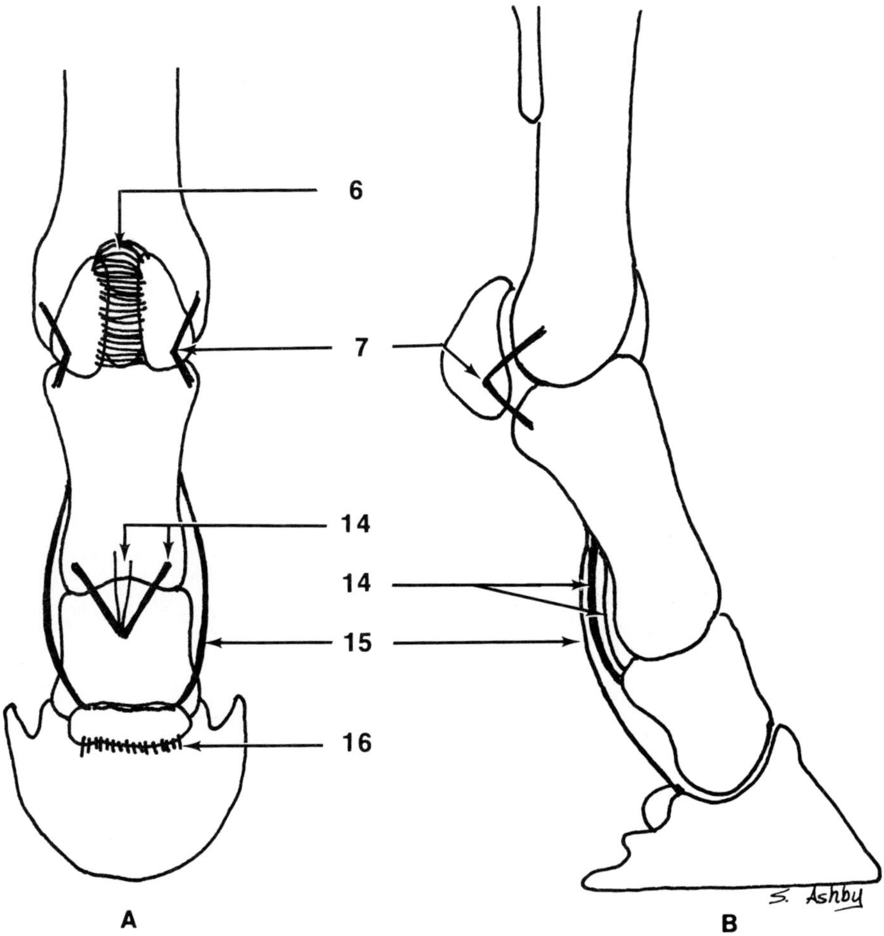

Figure 5. Ligaments of the equine thoracic limb (A-palmar aspect, B-lateral aspect)

6. Palmar ligaments of metacarpophalangeal joint
7. Collateral sesamoidean ligaments
14. Palmar ligaments of proximal interphalangeal joint
15. Collateral ligament of distal sesamoid
16. Distal sesamoidean ligament

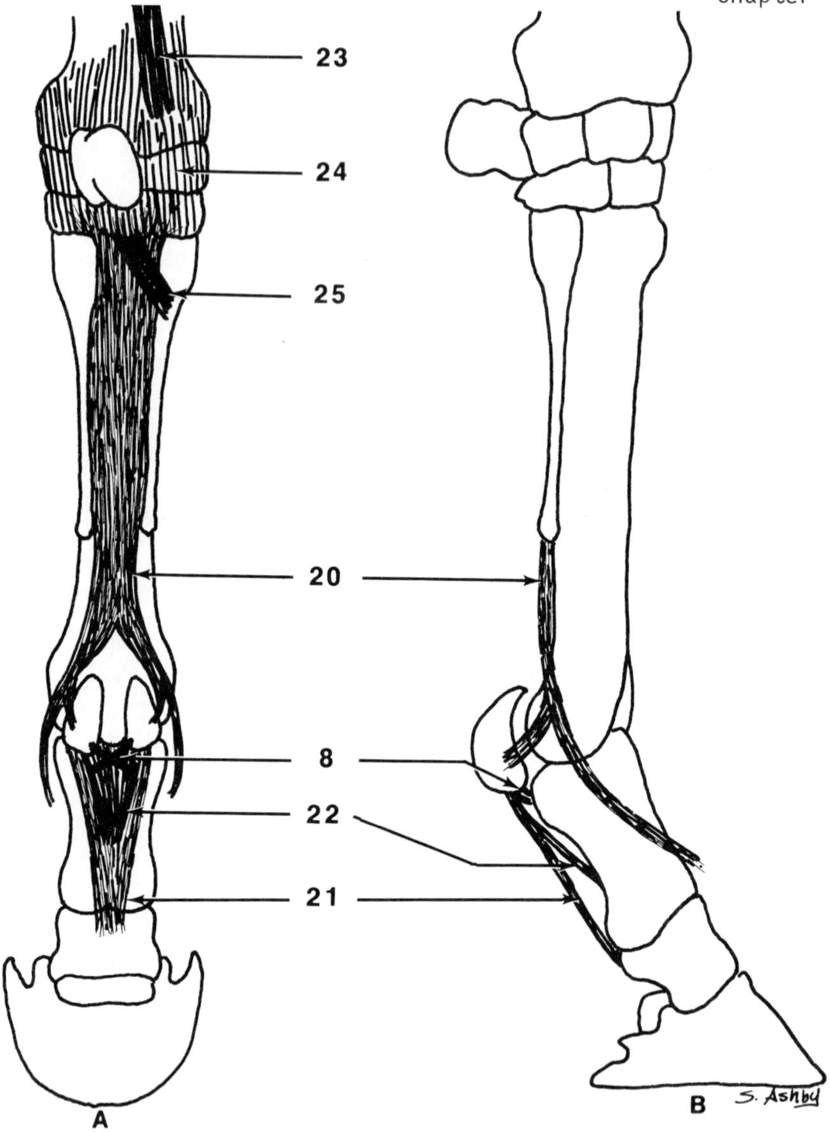

Figure 6. Suspensory apparatus of the equine metacarpophalangeal joint (A-palmar aspect, B-lateral aspect)

8. Cruciate sesamoidean ligament
20. Interosseous muscle
21. Straight sesamoidean ligament
22. Oblique sesamoidean ligament
23. Accessory ligament of superficial digital flexor muscle
24. Flexor retinaculum
25. Accessory ligament of deep digital flexor muscle

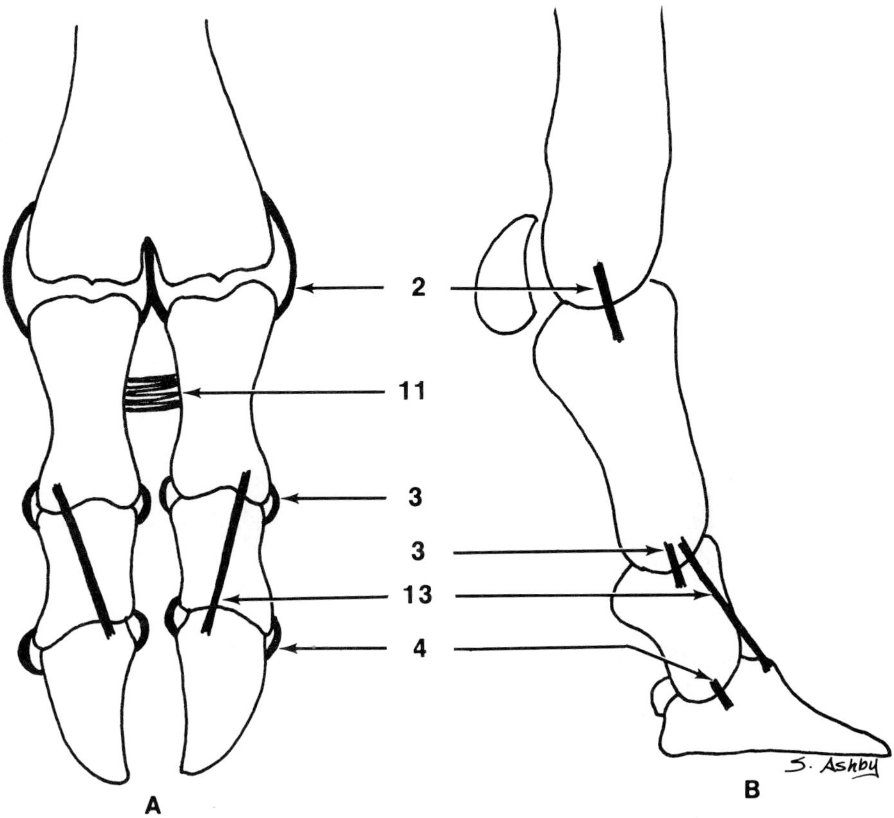

Figure 7. Ligaments of the bovine thoracic limb (A-dorsal aspect, B-lateral aspect)

2. Metacarpophalangeal collateral ligaments
3. Proximal interphalangeal collateral ligaments
4. Distal interphalangeal collateral ligaments
11. Proximal interdigital ligaments
13. Dorsal band

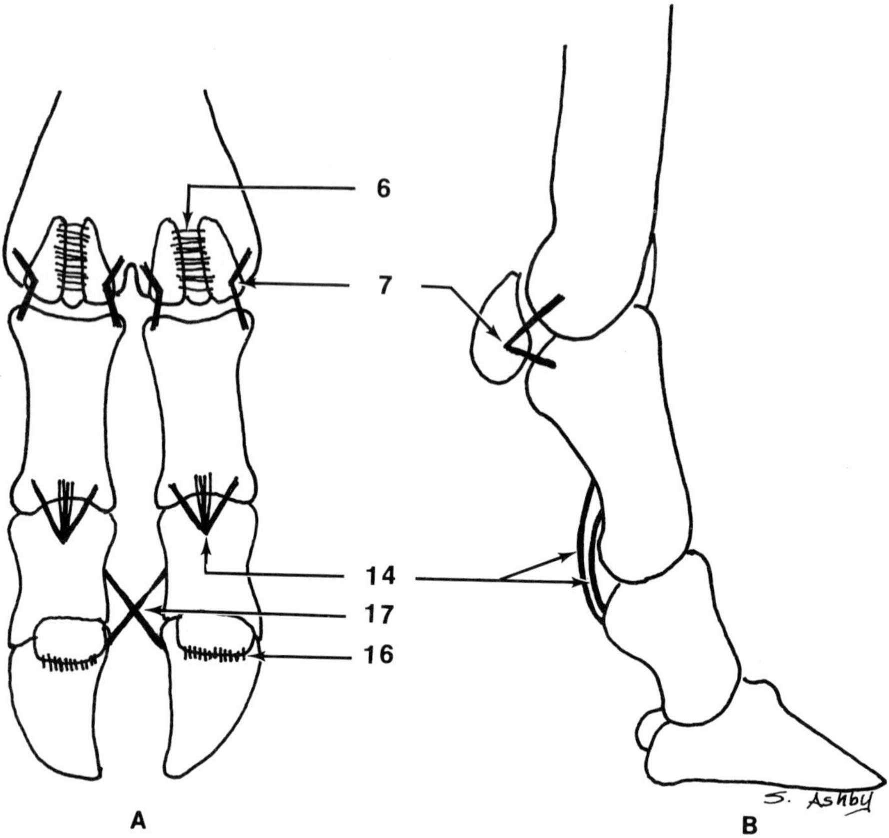

Figure 8. Ligaments of the bovine thoracic limb (A-palmar aspect, B-lateral aspect)

6. Palmar ligaments of metacarpophalangeal joint
7. Collateral sesamoidean ligaments
14. Palmar ligaments of proximal interphalangeal joint
16. Distal sesamoidean ligament
17. Distal interdigital ligaments

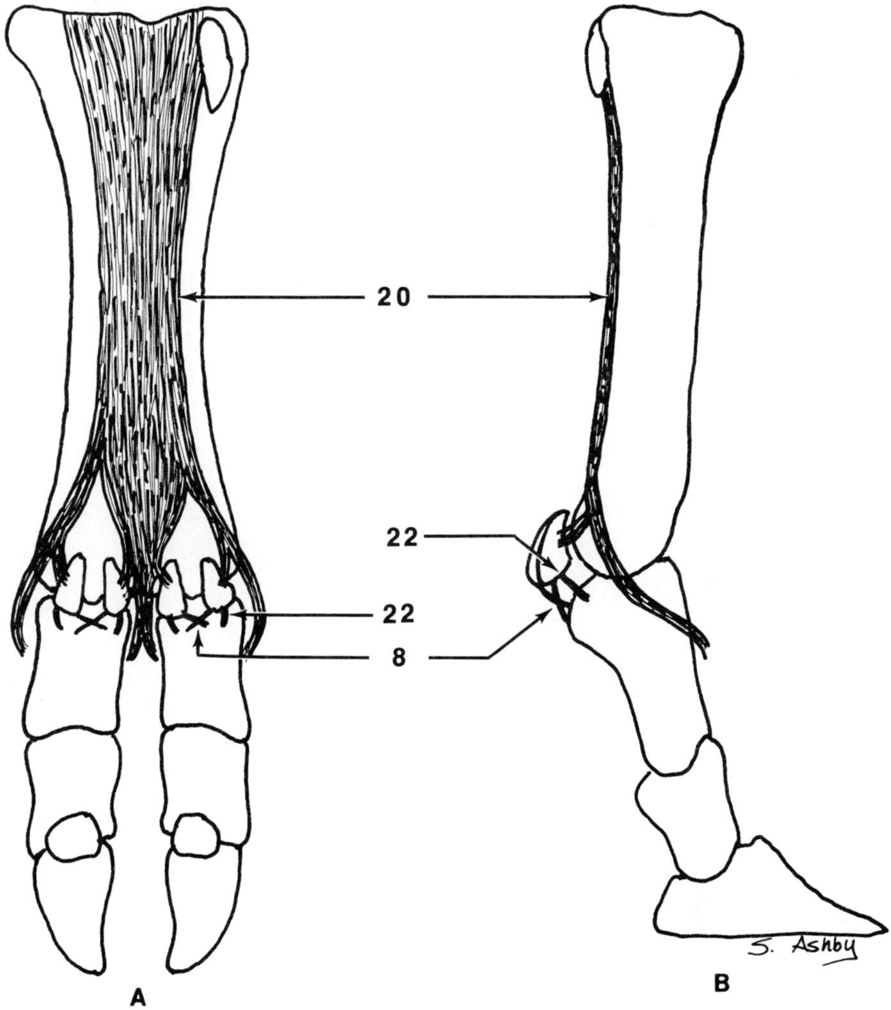

Figure 9. Suspensory apparatus of the bovine metacarpophalangeal joint (A-palmar aspect, B-lateral aspect)

8. Cruciate sesamoidean ligaments
20. Interosseous muscle
22. Oblique sesamoidean ligaments

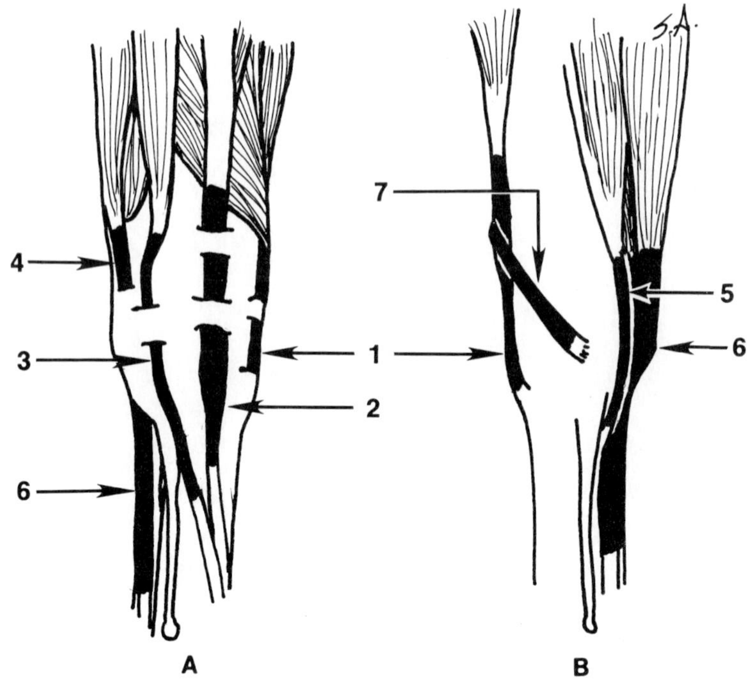

Figure 10. Synovial structures of the equine carpus (A-lateral aspect, B-medial aspect)

1. Extensor carpi radialis tendon sheath
2. Common digital extersor tendon sheath
3. Lateral digital extensor tendon sheath
4. Extensor carpi ulnaris tendon sheath
5. Flexor carpi radialis tendon sheath
6. Superficial and deep digital flexor tendon sheath
7. Abductor pollicis longus tendon sheath

JOINT CAPSULES

Humeral joint capsule
- Extends around biceps brachii tendon (--CF)
  - No bicipital bursa
- Injection sites
  - Insert needle cranioproximal to major tubercle and cranial to infraspinatus tendon; direct it ventromedially
  - Insert needle just cranial to acromial part of deltoideus m. (--CF)

Cubital joint capsule

Antebrachiocarpal joint capsule
- 1 of 2 highly movable carpal joints
- No communication with other carpal joints
- Injection sites
  - Medial to extensor carpi radialis tendon
  - Proximal to accessory carpal bone; between tendons of lateral digital extensor and extensor carpi ulnaris mm.

Middle carpal joint capsule
- 1 of 2 highly movable carpal joints
- Communicates with carpometacarpal joint
- Injection sites
  - Medial to extensor carpi radialis tendon

Carpometacarpal joint capsule
- Almost no movement
- Communicates with middle carpal joint
- Inject by way of middle carpal joint approach

Metacarpophalangeal joint capsule
- Communication of joint capsules for each digit (-B--)
  - Palmar aspect
- Injection sites
  - Dorsomedial to common digital extensor tendon
  - Proximal pouch of capsule extends proximally between interosseous m. and metacarpals for about 1/4 of metacarpus (EB--)

Proximal interphalangeal joint capsule
- Injection site
  - Dorsomedial to common digital extensor tendon

Distal interphalangeal joint capsule
- May communicate with (E---)
  - Navicular bursa
  - Deep digital flexor synovial sheath
- Injection site (EB--)
  - Insert needle 1.5 cm proximal to the coronary border of hoof at border of extensor tendon; direct needle obliquely distomedially

ARTERIES

Subclavian a.

Axillary a. (Figs. 11-14)
- External thoracic a.
  - Supplies
    - Pectoral mm.
- Lateral thoracic a.
  - Supplies
    - Lateral thorax
- Subscapular a.
  - Supplies
    - Caudal humeral area
    - Caudal scapular area
  - Thoracodorsal a.
    - Supplies
      - Latissimus dorsi m.
  - Caudal circumflex humeral a.
- Cranial circumflex humeral a.

Brachial a.
- Deep brachial a.
  - Supplies
    - Triceps brachii m.
- Collateral ulnar a.
  - Supplies
    - Cubital joint
    - Caudal antebrachial region
- Superficial brachial a. (--CF)
  - Supplies
    - Cranial, superficial cubital region
    - Superficial structures of antebrachium
  - Cranial superficial antebrachial a.
    - Located lateral to cephalic v.
    - Medial branch
      - Located medial to cephalic v.
    - Dorsal common digital aa.
      - Axial dorsal proper digital aa.
      - Abaxial dorsal proper digital aa.
- Transverse cubital a.
  - Supplies
    - Deep cubital region
  - Rete carpi dorsale (E---)
- Cranial interosseous a. (---F)
  - Passes between proximal radius and ulna
- Caudal interosseous a. (---F)
  - Located in interosseous space
  - Ulnar a.
    - Supplies
      - Deep digital flexor m.

Common interosseous a. (EBC-)
- Supplies
  - Deep structures of antebrachium
  - Structures of manus
- Ulnar a. (--C-)
  - Supplies
    - Deep digital flexor m.
- Cranial interosseous a.
  - Passes between proximal radius and ulna
  - Rete carpi dorsale (EB--)
- Caudal interosseous a.
  - Located in interosseous space (--C-)

Median a.
- Major artery to manus (EBC-)
- Deep antebrachial a.
- Proximal radial a. (E---)
  - Rete carpi dorsale
    - Dorsal metacarpal aa.
- Radial a.
  - Major artery to manus (---F)
  - Rete carpi dorsale (-BCF)
    - Dorsal metacarpal aa.
  - Deep palmar arch
    - Palmar metacarpal aa.
- Palmar branch (E---)
  - Palmar common digital a. III
- Superficial palmar arch (eBCF)
  - Palmar common digital aa. (-BCF)
    - Palmar proper digital aa.

Palmar common digital a. II (E---)
- Palmar proper digital aa.

(Note: Superficial vessels and nerves are called common digital in the region of the metacarpus and proper digital in the region of the phalanges. Deep vessels and nerves are called metacarpal.)

VEINS

Major veins follow the arteries (Figs. 15-18)
- Medial and lateral palmar (proper digital) vv. (E---)
  - Located dorsal to artery and nerve

Cephalic v. (Figs. 15-18)
- Joins external jugular v.
- Location
  - In a groove between descending pectoral and cleidobrachialis mm. (EB--)
  - Crosses lateral aspect of brachiocephalicus m. (EB--)
  - Crosses medial aspect of brachiocephalicus m. (--CF)

1. Axillary a.
2. External thoracic a.
3. Lateral thoracic a.
4. Subscapular a.
5. Thoracodorsal a.
6. Caudal circumflex humeral a.
7. Cranial circumflex humeral a.
8. Deep brachial a.
9. Brachial a.
10. Collateral ulnar a.
14. Transverse cubital a.
15. Common interosseous a.
17. Cranial interosseous a.
19. Deep antebrachial a.
20. Median a.
21. Radial a.
29. Lateral palmar metacarpal a.
30. Medial palmar metacarpal a.
31. Lateral dorsal metacarpal a. (III)
32. Medial dorsal metacarpal a. (II)
33. Medial (palmar proper) digital a. (II)
34. Lateral (palmar proper) digital a. (III)
37. Medial palmar (common digital) a. (II)
38. Lateral palmar (common digital) a. (III)
39. Proximal radial a.

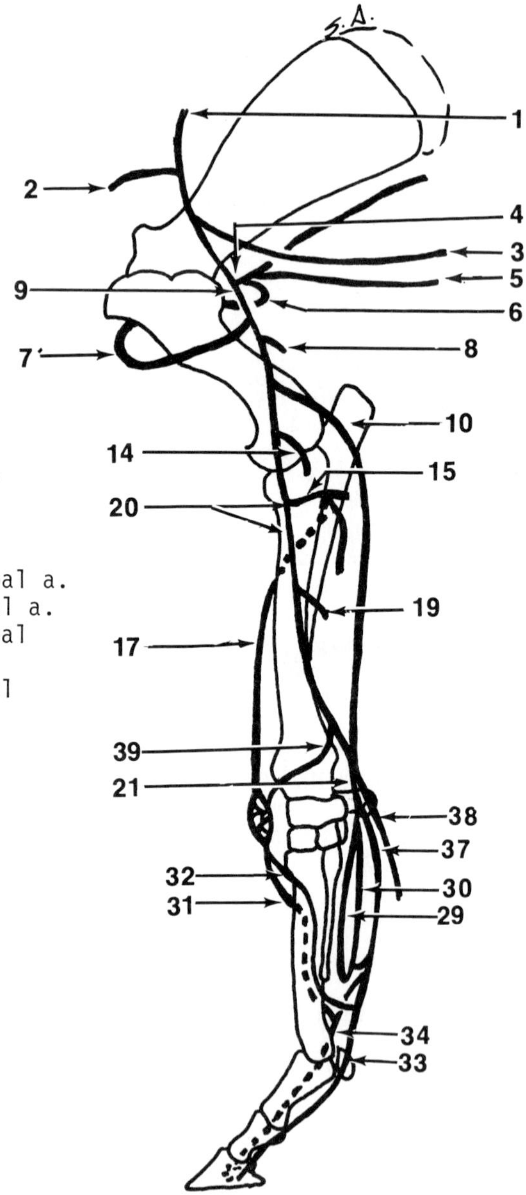

Figure 11. Thoracic limb arteries of the horse (medial aspect)

1. Axillary a.
2. External thoracic a.
3. Lateral thoracic a.
4. Subscapular a.
5. Thoracodorsal a.
6. Caudal circumflex humeral a.
7. Cranial circumflex humeral a.
8. Deep brachial a.
9. Brachial a.
10. Collateral ulnar a.
14. Transverse cubital a.
15. Common interosseous a.
17. Cranial interosseous a.
19. Deep antebrachial a.
20. Median a.
21. Radial a.
29. Lateral palmar metacarpal a.
30. Medial palmar metacarpal a.
31. Dorsal metacarpal a. III
33. Palmar proper digital a. III
34. Palmar proper digital a. IV
35. Dorsal proper digital a. III
36. Dorsal proper digital a. IV
38. Palmar common digital a. III
40. Superficial palmar arch

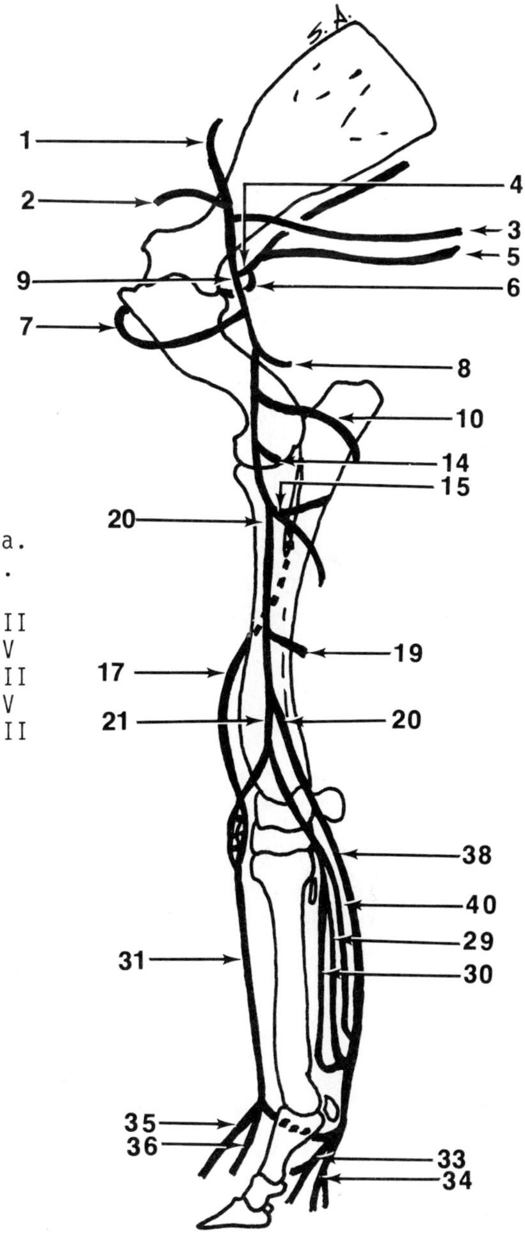

Figure 12. Thoracic limb arteries of cattle (medial aspect)

1. Axillary a.
2. External thoracic a.
3. Lateral thoracic a.
4. Subscapular a.
5. Thoracodorsal a.
6. Caudal circumflex humeral a.
7. Cranial circumflex humeral a.
8. Deep brachial a.
9. Brachial a.
10. Collateral ulnar a.
14. Transverse cubital a.
15. Common interosseous a.
17. Cranial interosseous a.
19. Deep antebrachial a.
20. Median a.
21. Radial a.
22. Palmar common digital aa.
23. Palmar metacarpal aa.
24. Palmar proper digital aa.
26. Dorsal common digital aa.
27. Dorsal proper digital aa.

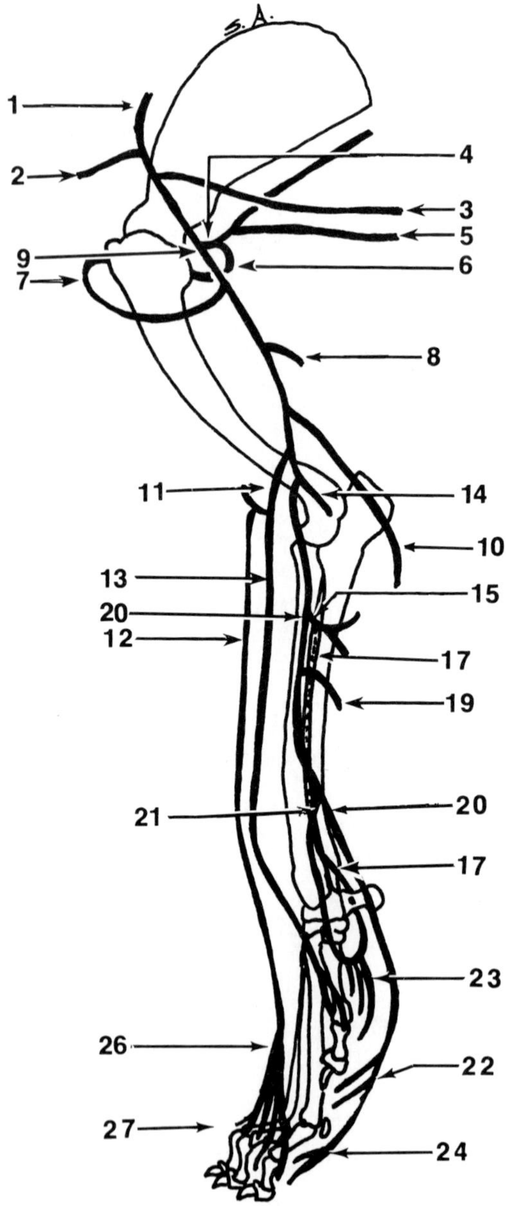

Figure 13. Thoracic limb arteries of the dog (medial aspect)

1. Axillary a.
2. External thoracic a.
3. Lateral thoracic a.
4. Subscapular a.
5. Thoracodorsal a.
6. Caudal circumflex humeral a.
7. Cranial circumflex humeral a.
8. Deep brachial a.
9. Brachial a.
10. Collateral ulnar a.
14. Transverse cubital a.
16. Ulnar a.
17. Cranial interosseous a.
19. Deep antebrachial a.
20. Median a.
21. Radial a.
22. Palmar common digital aa.
23. Palmar metacarpal aa.
24. Palmar proper digital aa.
26. Dorsal common digital aa.
27. Dorsal proper digital aa.
31. Dorsal metacarpal aa.

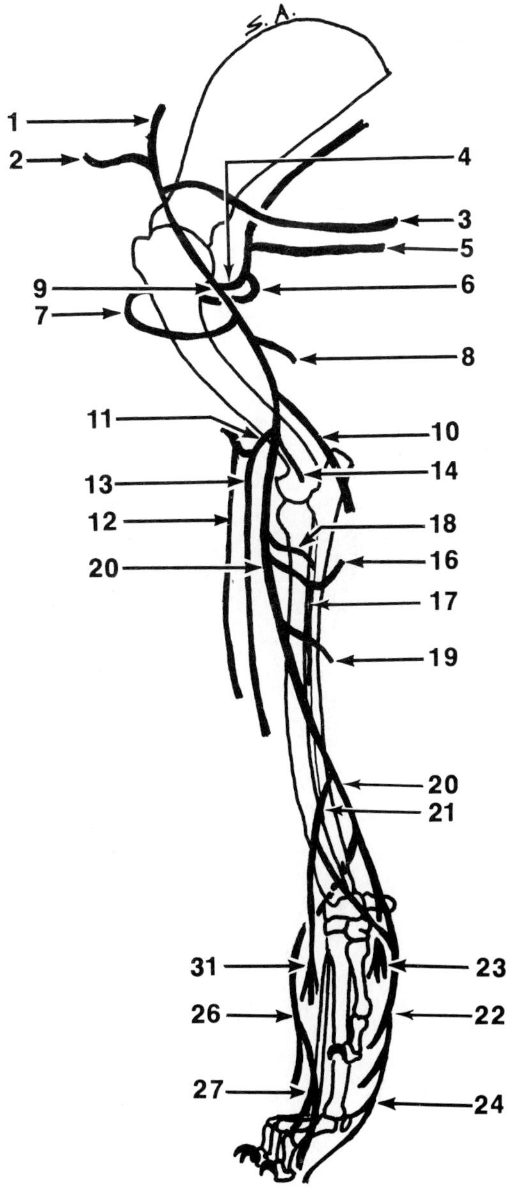

Figure 14. Thoracic limb arteries of the cat (medial aspect)

Cephalic v., location (cont.)

- Covered by brachioradialis m. in antebrachium (--CF)
- Arises from palmar manus

Axillobrachial v. (--CF)

- Joins axillary v.
- Location
  - Goes medial to deltoideus m.
  - Accompanies
    - Axillary n.
    - Caudal circumflex humeral a. & v.
- Omobrachial v. (--C-)
  - Joins external jugular v.
  - Location
    - Crosses lateral aspect of brachiocephalicus m.

Median cubital v.

- Joins brachial v.

Accessory cephalic v.

- Location
  - Craniomedial antebrachium (EB--)
  - Cranial antebrachium (--CF)
- Arises from dorsal manus

NERVES

Suprascapular n.

- Spinal nerve contributions
  - C_5 (--c-)
  - C_6
  - C_7 (EBCf)
- Supplies motor fibers to
  - Supraspinatus m.
  - Infraspinatus m.
- Course of travel
  - Goes laterally between supraspinatus and subscapularis mm.
  - Crosses cranial border and lateral side of scapula ventral to scapular spine

Subscapularis n.

- Spinal nerve contributions
  - C_6 (eBCF)
  - C_7
  - C_8 (---f)
- Supplies motor fibers to
  - Subscapularis m.

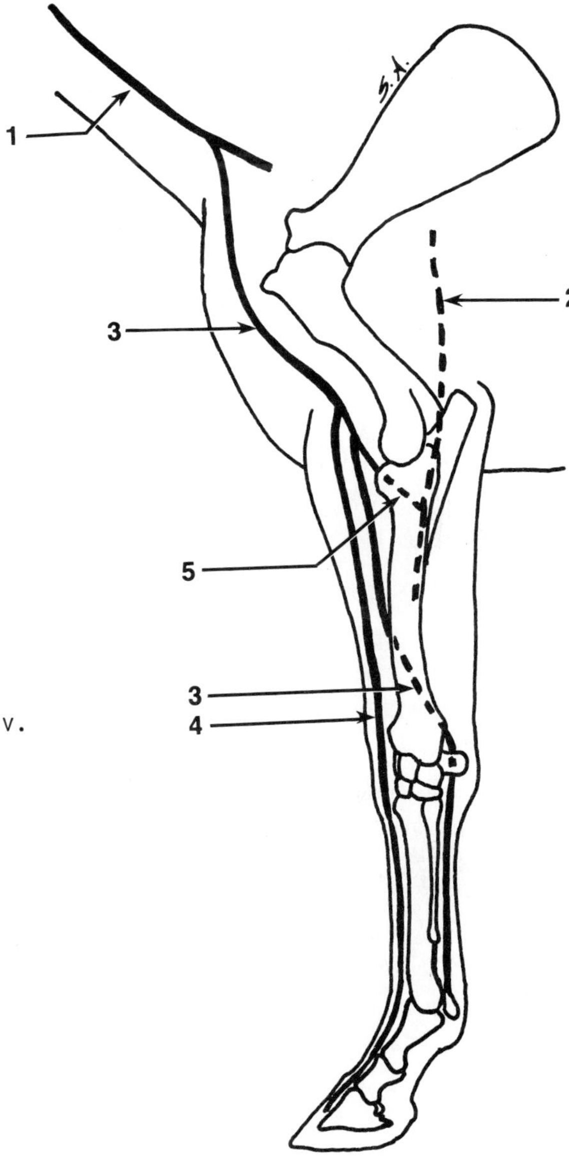

1. External jugular v.
2. Brachial v.
3. Cephalic v.
4. Accessory cephalic v.
5. Median cubital v.

Figure 15. Thoracic limb veins of the horse (lateral aspect)

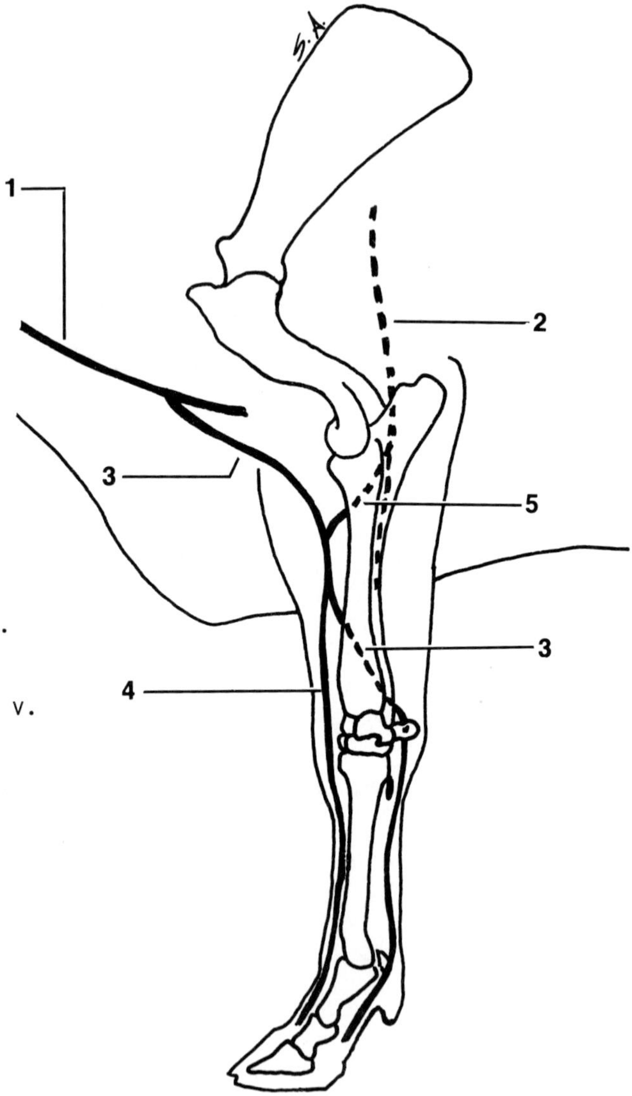

1. External jugular v.
2. Brachial v.
3. Cephalic v.
4. Accessory cephalic v.
5. Median cubital v.

Figure 16. Thoracic limb veins of cattle (lateral aspect)

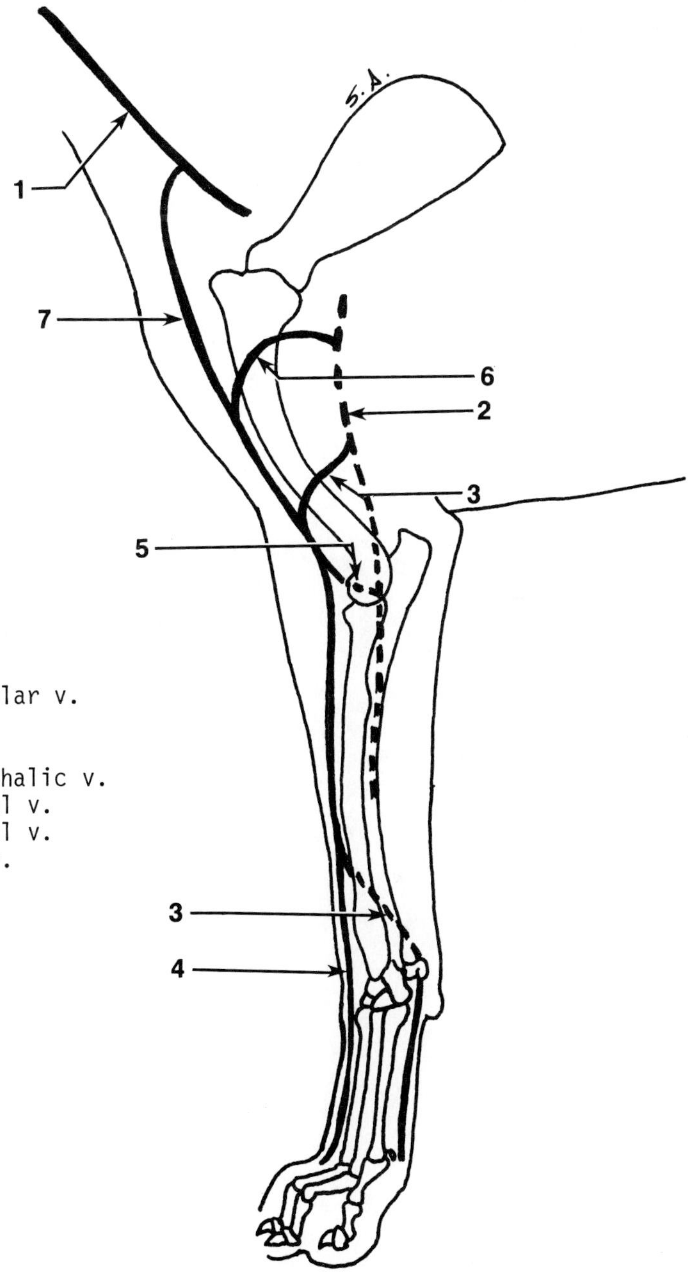

1. External jugular v.
2. Brachial v.
3. Cephalic v.
4. Accessory cephalic v.
5. Median cubital v.
6. Axillobrachial v.
7. Omobrachial v.

Figure 17. Thoracic limb veins of the dog (lateral aspect)

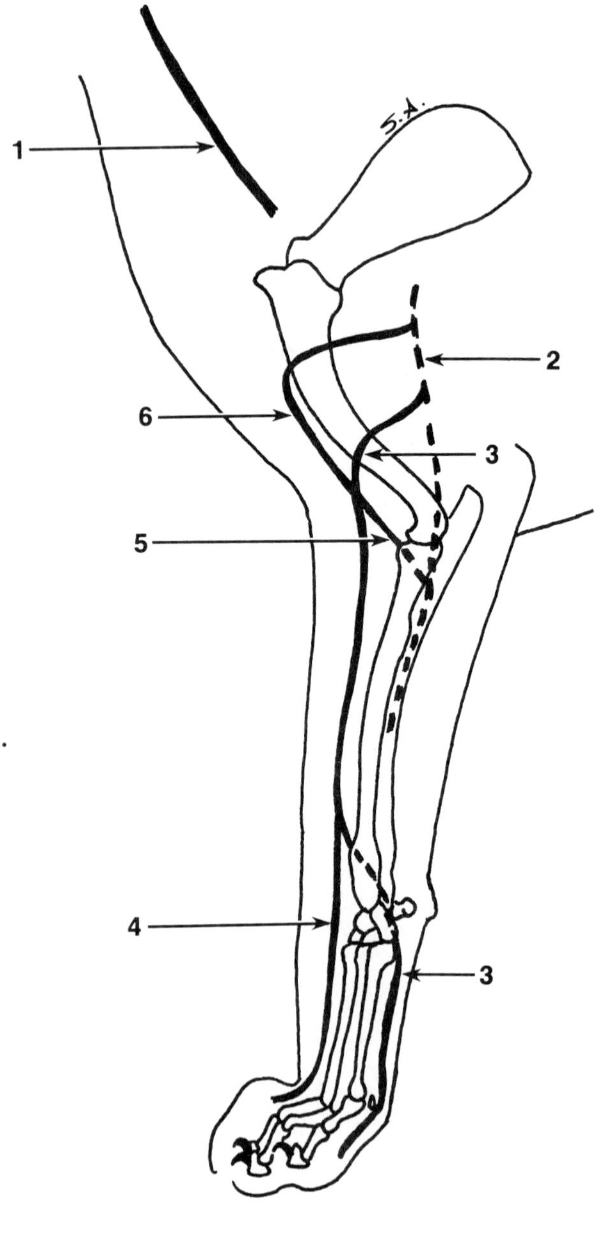

1. External jugular v.
2. Brachial v.
3. Cephalic v.
4. Accessory cephalic v.
5. Median cubital v.
6. Axillobrachial v.

Figure 18. Thoracic limb veins of the cat (lateral aspect)

Musculocutaneous n.
- Spinal nerve contributions
  - C_6 (-BCF)
  - C_7
  - C_8 (EBC-)
- Supplies motor fibers to
  - Flexors of cubital joint
- Sensory to (Figs. 19,20)
  - Medial cutaneous antebrachium
- Course of travel
  - Sends branch through coracobrachialis m. to the biceps brachii m. (EB--)
  - Joins the median n. near proximal humerus and separates from it near elbow (EB--)
  - Continues as the medial cutaneous antebrachial n.
    - Emerges between distal ends of biceps brachii and brachialis mm.
    - Crosses lacertus fibrosus from lateral to medial (E---)
    - Continues to
      - Carpus (-B--)
      - Metacarpophalangeal joint with cephalic and accessory cephalic vv. (E-CF)

Axillary n.
- Spinal nerve contributions
  - C_6 (e-cF)
  - C_7
  - C8 (EBC-)
- Supplies motor fibers to
  - Flexors of humeral joint
- Sensory to (Figs. 19,20)
  - Lateral brachium
  - Proximolateral antebrachium
- Courses of travel
  - Laterally between subscapularis and teres major mm.
    - Travels with
      - Caudal circumflex humeral a.
      - Axillobrachial v. (--CF)
    - Continues as
      - Cranial cutaneous antebrachial n.
      - Lateral cutaneous brachial n.

Thoracodorsal n.
- Spinal nerve contributions
  - C_7 (-BcF)
  - C_8
  - T_1 (ebc-)
- Supplies motor fibers to
  - Latissimus dorsi m.

Radial n.
- Spinal nerve contributions
  - C_6 (---f)
  - C_7 (eBCF)
  - C_8
  - T_1
  - T_2 (--c-)
- Supplies motor fibers to
  - Extensors of cubital joint
  - Extensors of carpus and digits
- Sensory to (Figs. 19,20)
  - Lateral antebrachium
  - Cranial antebrachium (-BCF)
  - Cranial manus (-BCF)
- Course of travel (Figs. 21-28)
  - Curves caudally around humerus, next to the bone, from medial to lateral
    - Sends large branch to triceps brachii m.
  - Deep branch
    - Crosses elbow deep to the supinator m. (--CF)
    - Sends branches to carpal and digital extensors
  - Superficial branch (-BCF)
    - Divides into 2 branches (--CF)
      - Located on either side of cephalic v.
    - Communicates at carpus with medial cutaneous antebrachial n. of the musculocutaneous n. (-B--)
    - Divides into 3 branches (-B--)
      - Axial dorsal common digital n. of III
      - Axial dorsal common digital n. of IV
      - Abaxial dorsal common digital n. of III

Median n.
- Spinal nerve contributions
  - C_7 (e--F)
  - C_8
  - T_1
  - T_2 (EBc-)
- Supplies motor fibers to
  - Flexors of carpus and digits
  - Pronator mm. (--CF)
- Sensory to (Figs. 19,20)
  - Mediopalmar manus
- Course of travel (Figs. 21-28)
  - Located caudal to brachial a.
  - Joined by branch of musculocutaneous n.
    - Near proximal humerus (EB--)
    - Near distal humerus (--CF)

Sends an anastomotic branch to the ulnar n. near the carpus (EB--)
Passes under pronator teres m. (--CF)
Becomes the (medial) palmar common digital n. (E-III)
Sends an anastomotic branch caudal to flexor tendons to the lateral palmar (common digital) n. (E---)
Divides near the metacarpophalangeal joint
- Medial dorsal (proper) digital n. (E---)
- Medial palmar (proper) digital n. (E---)
- Axial palmar (proper) digital n. I (--CF)
- Abaxial palmar (proper) digital n. II (--CF)
- Axial palmar (proper) digital n. II (--CF)
- Abaxial palmar (proper) digital n. III (-BCF)
- Axial palmar (proper) digital n. III (-BCF)
- Abaxial palmar (proper) digital n. IV (--CF)
- Communicating branch to the ulnar n. to form the abaxial palmar (proper) digital n. IV

Ulnar n.
- Spinal nerve contributions
  - C_8 (-BCF)
  - T_1
  - T_2 (EBc-)
- Supplies motor fibers to
  - Flexors of carpus and digits
- Sensory to (Figs 19,20)
  - Caudolateral antebrachium
  - Dorsolateral manus
  - Lateropalmar manus
- Course of travel (Figs. 21-28)
  - Lies caudal to brachial a.
  - Crosses medial epicondyle of humerus
  - Gives rise to the caudal cutaneous antebrachial n.
  - Passes deep to flexor carpi ulnaris and ulnaris lateralis tendons
  - Gives rise to 2 branches
    - Dorsal branch
      - Cutaneous to dorsolateral manus
    - Palmar branch
      - Deep branch
        - Motor and sensory to palmar structures as the palmar metacarpal nn.
      - Superficial branch
        - Unites with a branch from the median n.
          - Abaxial palmar proper digital n. IV (-BCF)
          - Abaxial palmar proper digital n. V (--CF)
          - Axial palmar (proper) digital n. V(--CF)
          - Lateral palmar (proper) digital n. (E---)
          - Medial and lateral palmar metacarpal nn. (E---)

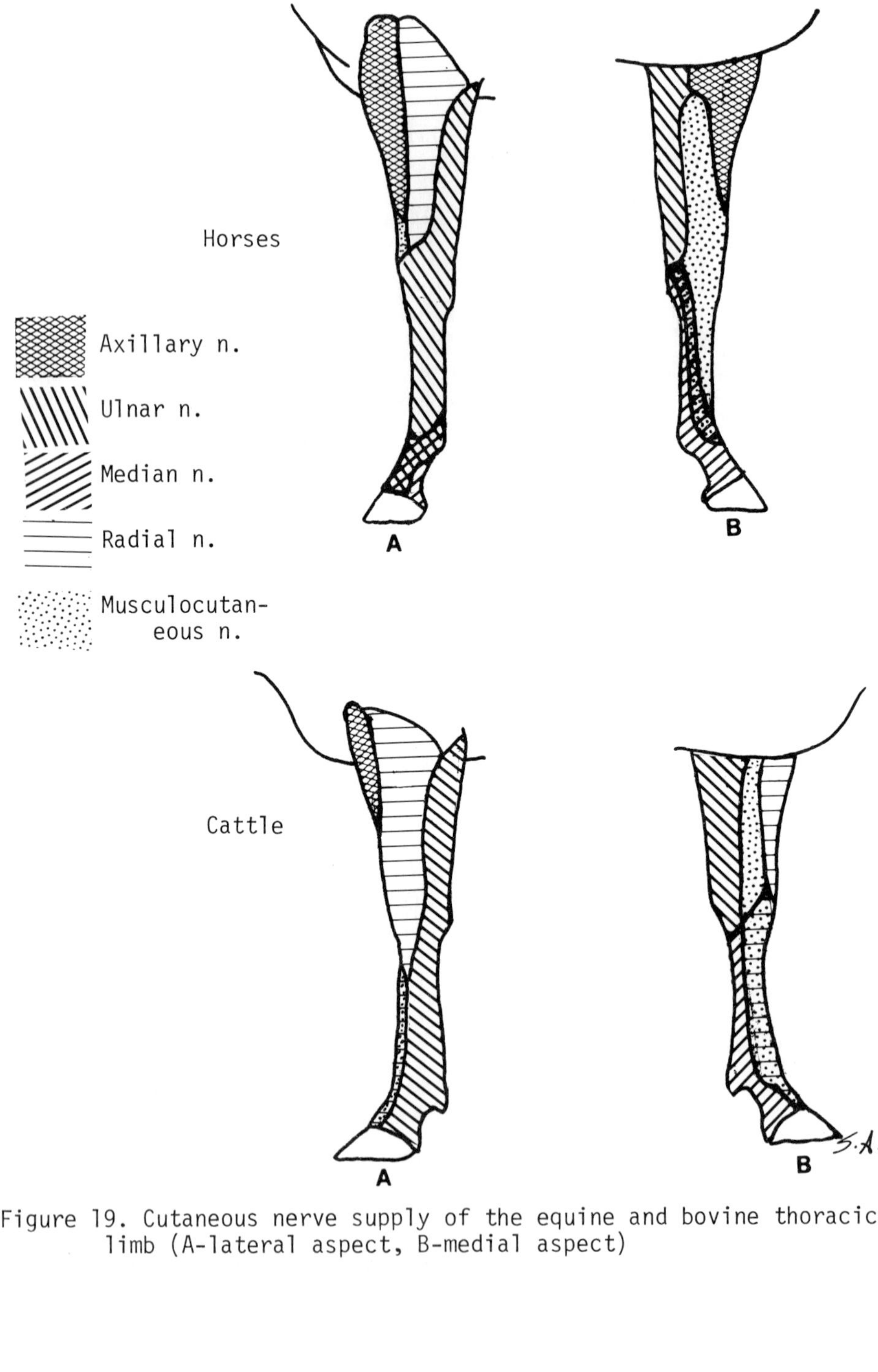

Figure 19. Cutaneous nerve supply of the equine and bovine thoracic limb (A-lateral aspect, B-medial aspect)

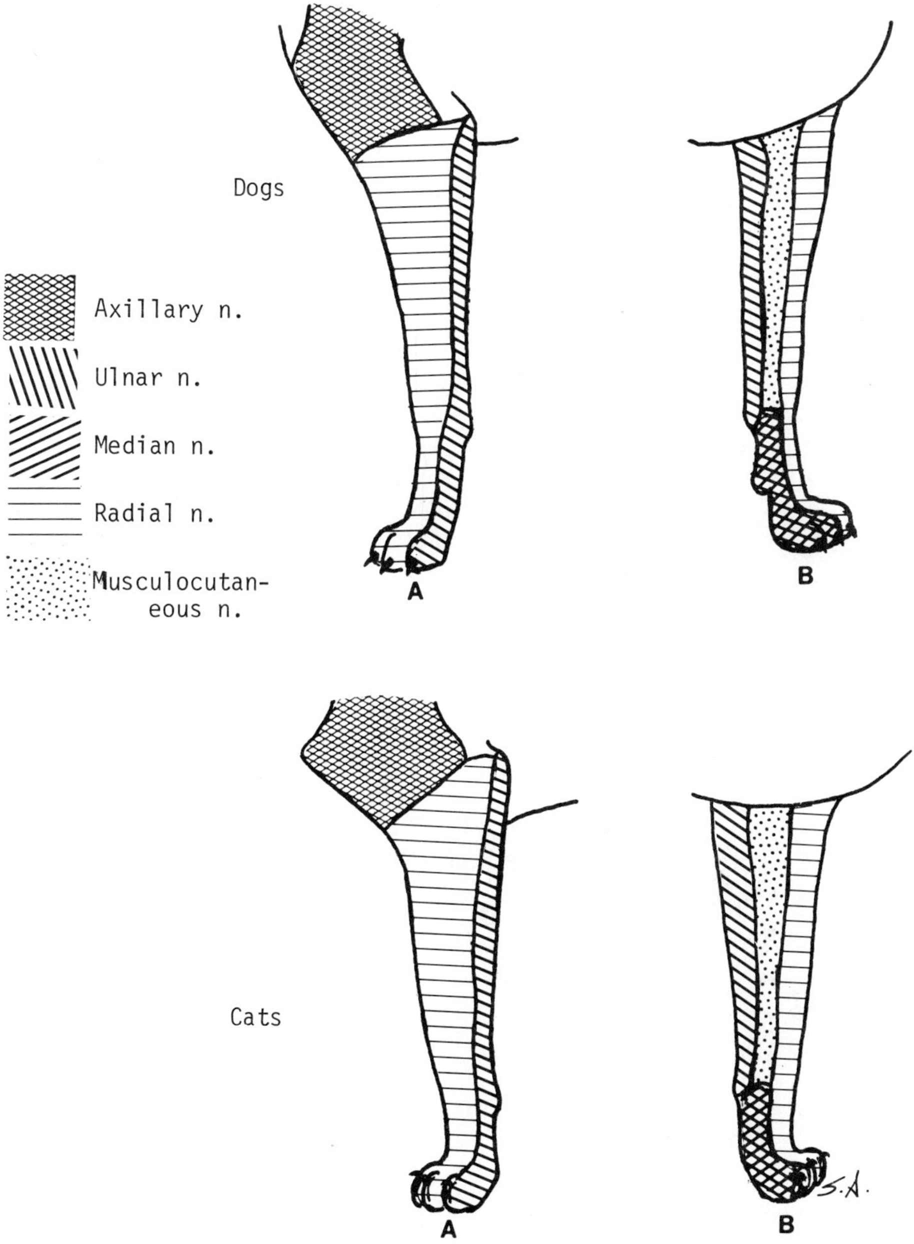

Figure 20. Cutaneous nerve supply of the canine and feline thoracic limb (A-lateral aspect, B-medial aspect)

1. Median n.
2. Ulnar n.
3. Dorsal branch
4. Palmar branch
9. Medial palmar (common digital) n.
10. Lateral palmar (common digital) n.
12. Deep branch
13. Medial palmar metacarpal n.
14. Lateral palmar metacarpal n.
15. Communicating branch
16. Medial palmar (proper) digital n.
17. Medial dorsal (proper) digital n.
18. Lateral palmar (proper) digital n.
19. Lateral dorsal (proper) digital n.

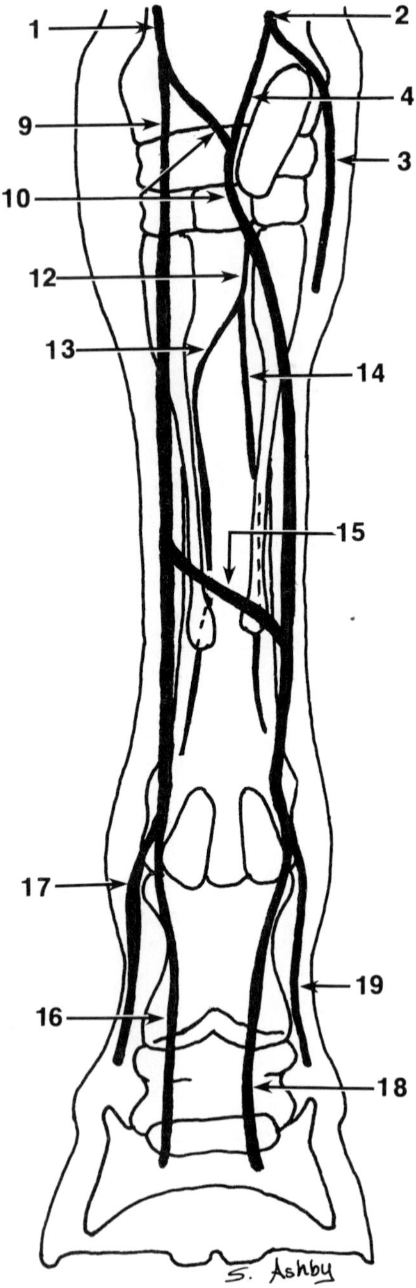

Figure 21. Nerves of the equine manus (right limb, palmar aspect)

2. Ulnar n.
3. Dorsal branch
10. Lateral palmar (common digital) n.
14. Lateral palmar metacarpal n.
15. Communicating branch
18. Lateral palmar (proper) digital n.
19. Lateral dorsal (proper) digital n.

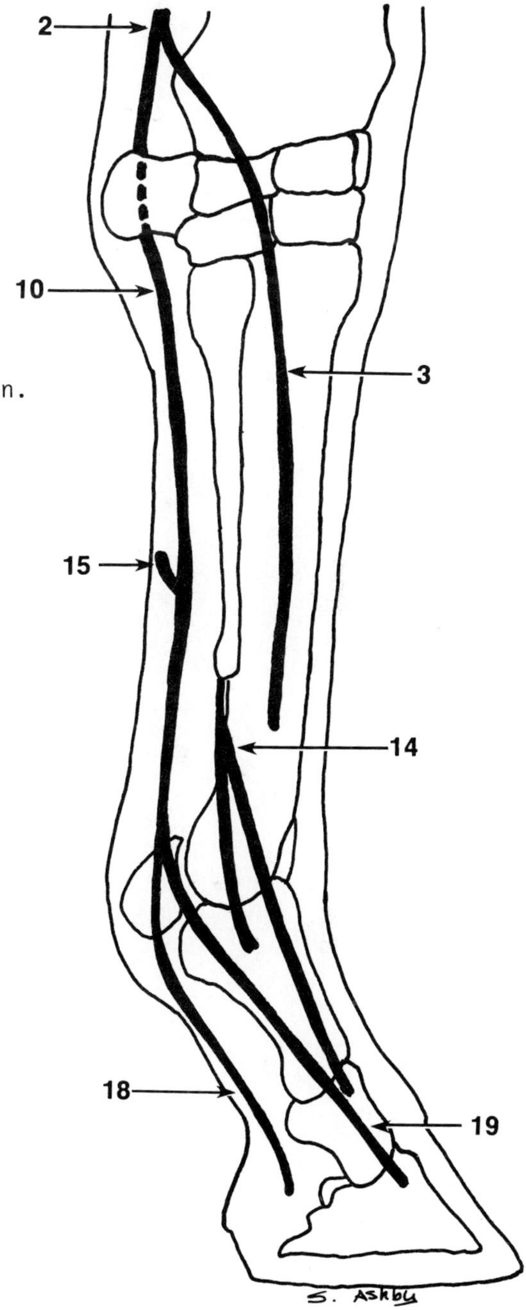

Figure 22. Nerves of the equine manus (right limb, lateral aspect)

1. Median n.
5. Medial cutaneous antebrachial n. (of the musculocutaneous n.)
9. Medial palmar (common digital) n.
13. Medial palmar metacarpal n.
15. Communicating branch
16. Medial palmar (proper) digital n.
17. Medial dorsal (proper) digital n.

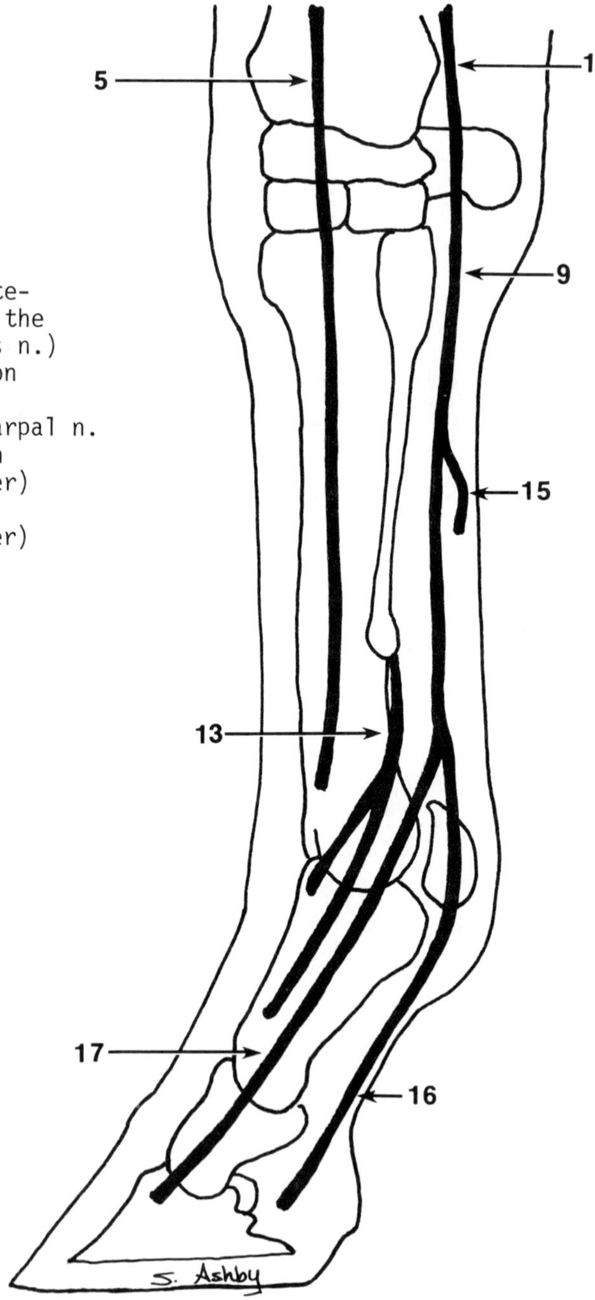

Figure 23. Nerves of the equine manus (right limb, medial aspect)

1. Median n.
2. Ulnar n.
3. Dorsal branch
9. Medial palmar common digital n.
10. Lateral palmar common digital n.
20. Axial palmar proper digital n. III
21. Abaxial palmar proper digital n. III
22. Axial palmar proper digital n. IV
23. Abaxial palmar proper digital n. IV

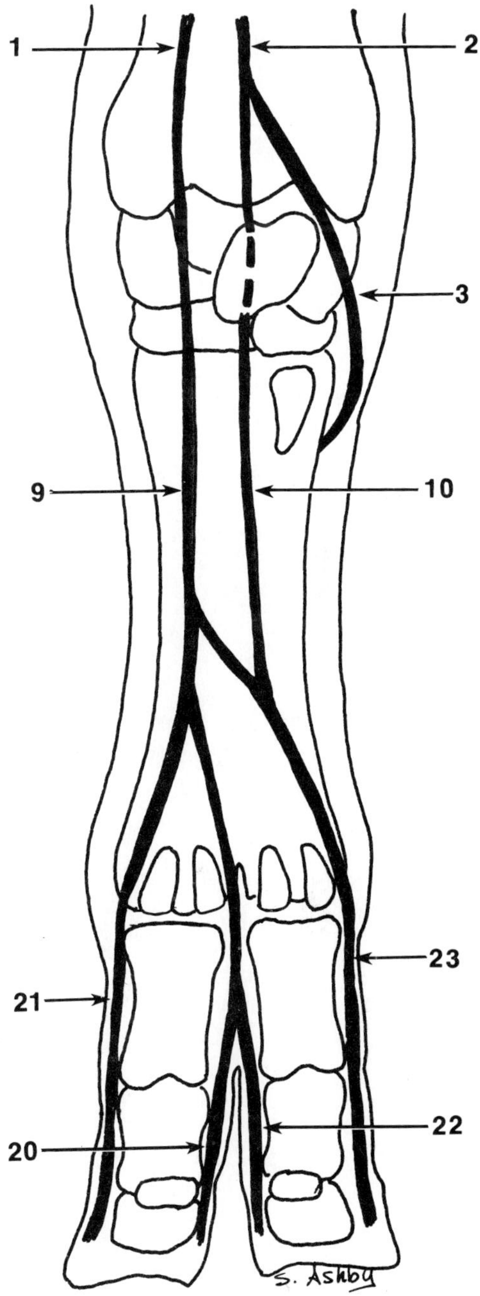

Figure 24. Nerves of the bovine manus (right limb, palmar aspect)

3. Dorsal branch (ulnar n.)
4. Palmar branch (ulnar n.)
9. Medial palmar common digital n.
10. Lateral palmar common digital n.
22. Axial palmar proper digital n. IV
23. Abaxial palmar proper digital n. IV
25. Lateral dorsal common digital n.
28. Axial dorsal proper digital n. III
29. Axial dorsal proper digital n. IV
30. Abaxial dorsal proper digital n. IV

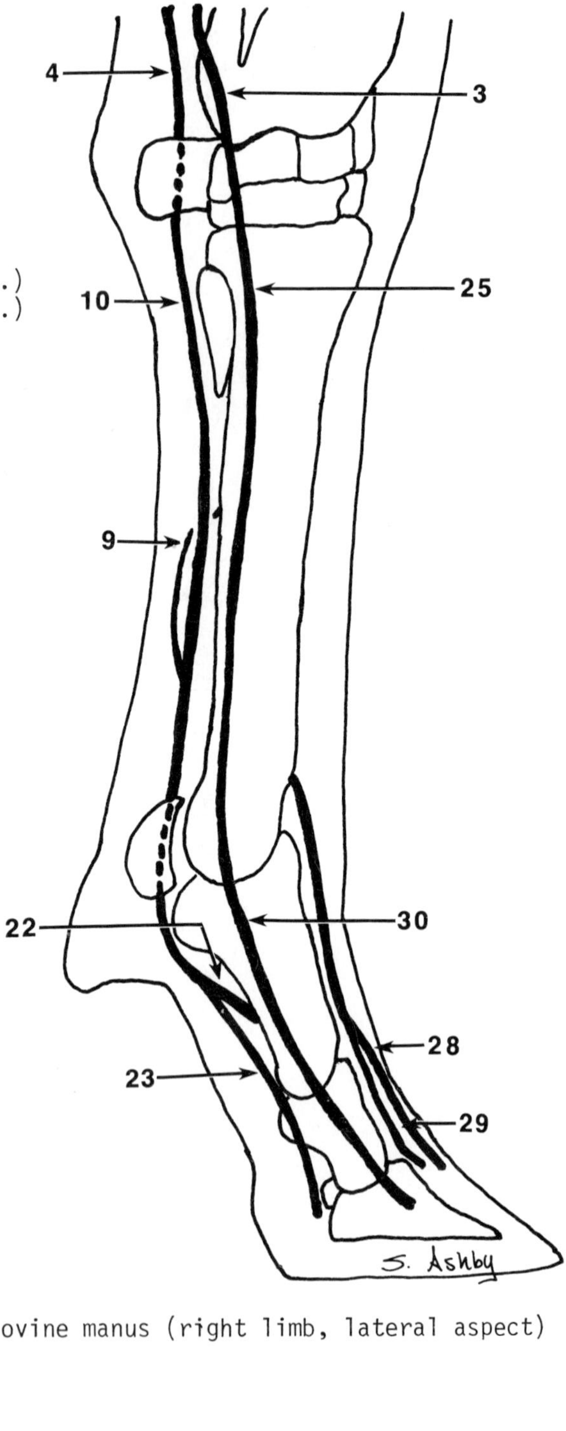

Figure 25. Nerves of the bovine manus (right limb, lateral aspect)

5. Medial cutaneous antebrachial n. (of the musculocutaneous n.)
6. Superficial branch (radial n.)
26. Medial dorsal common digital n.
28. Axial dorsal proper digital n. III
29. Axial dorsal proper digital n. IV
31. Abaxial dorsal proper digital n. III

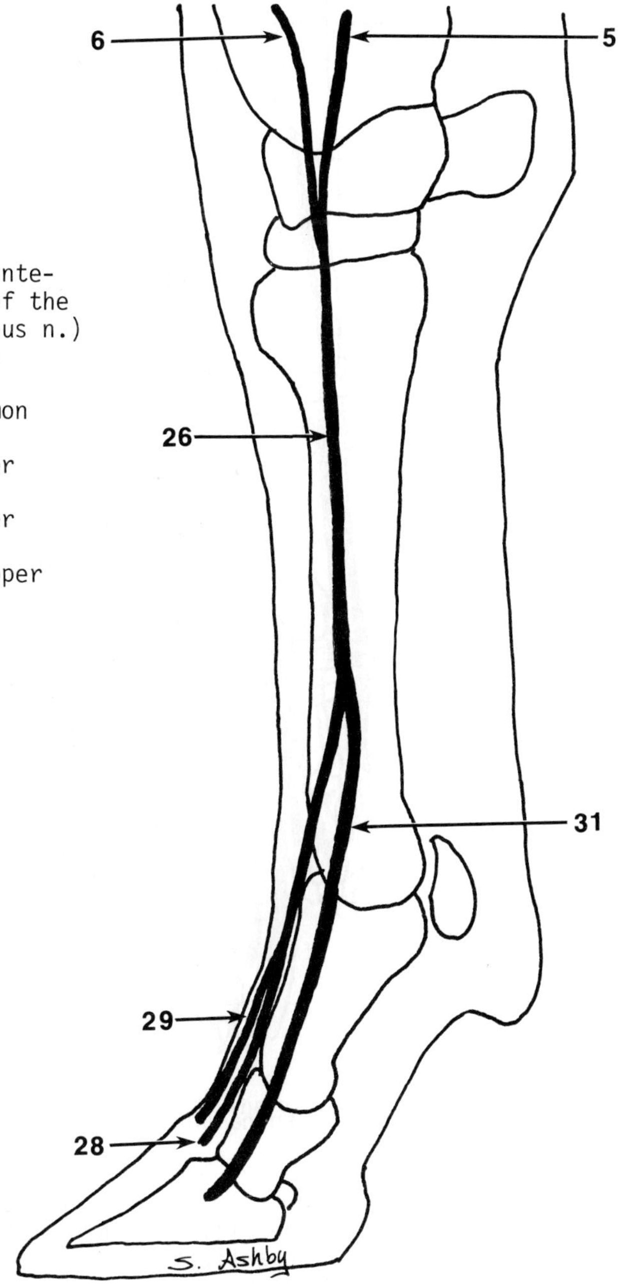

Figure 26. Nerves of the bovine manus (right limb, medial aspect)

Internal structures, ungual cartilages (cont.)
Function
Part of blood return mechanism

PALPATION

Rhomboideus cervicis m. (EB--)
Not to be confused with crest (E-)
Forms Brahman bull hump (-B)
Spinous processes of T_{2-4} (E---)
Location of supraspinous bursa
Highest spinous process is T_4 or T_5
Point of the withers
Serratus ventralis (cervical portion) (E---)
Scapula
Spine
Acromion (--CF)
Hamate process (---F-)
Suprahamate process (---F-)
Supraglenoid tubercle (--CF)
Supraspinatus m.
Infraspinatus m.
Tendon of termination
Superficial cervical ln.
Subclavius m. (E---)
Humerus
Major tubercle
Minor tubercle (--CF)
Deltoid tuberosity
Medial and lateral epicondyles
Deltoideus m. (acromial part) (--CF)
Biceps brachii m.
Tendon of origin (-BCF)
Termination (E---)
Lacertus fibrosus (E---)
Brachialis m.
Triceps brachii m.
Long head
Lateral head
Termination
Brachial a.
As it crosses the elbow between flexor carpi radialis m. and medial epicondyle of humerus
Radius and ulna
Olecranon
Medial and lateral styloid processes
Lateral collateral lig. of the cubital joint (--CF)
Cephalic v.
Accessory cephalic v.
Median a. (E---)
In groove between radius and flexor carpi radialis m.
Through transverse pectoral m.

Muscles of the forearm and their tendons (may need to raise and move a foot)
- Extensor carpi radialis m.
- Abductor pollicis longus m.
- Common digital extensor m.
- Lateral digital extensor m. (EB--)
- Ulnaris lateralis m.
- Flexor carpi ulnaris m.
- Superficial digital flexor m.
  - 2 tendons (-B--)
- Deep digital flexor m.
- Flexor carpi radialis m.

Carpus
- Accessory carpal bone
- Medial and lateral collateral carpal ligg.
- Carpal joint spaces

Torus carpeus (E-CF)
- Chestnut (E---)
- Carpal pad (--CF)
- (Note position relative to carpus)

Metacarpal bones

Interosseous m. (suspensory lig.)
- Branch to sesamoid bones (E---)
- Branches to digital extensor tendons (E---)

Metacarpophalangeal joint
- Proximal (palmar) sesamoid bones (EB--)
- Medial and lateral collateral sesamoid ligg. (E---)
  (Pick up foot to relax interosseous m.)
- Proximal pouch of joint capsule (E---)
- Calcar metacarpeum (E-CF)
  - Ergot (E---)
  - Metacarpal pad (--CF)

Medial and lateral (palmar proper) digital a. (E---)

Medial and lateral palmar (common digital) nn. (E---)
- Communicating branch

Proximal phalanx

Ungual cartilages (E---)

Digital annular lig. (E---)
- Proximal edge of distal part can be palpated as it crosses deep digital flexor tendon
  - This is lowest point of access to digital synovial sheath of the flexor tendons

Oblique sesamoidean ligg. (E---)

Distal lig. of the dewclaw (-B--)

Proximal interphalangeal joint
- Collateral ligg. of the joint (EB--)

Distal interphalangeal joint (--CF)
- Torus digitalis (-BCF)
  - Torus ungulae (dewclaw) (-B--)
  - Digital pads (--CF)

Structures of the foot (EB--)
- Coronary band (EB--)

Structures of the foot (cont.)
- Periople (EB--)
- Hoof wall (EB--)
- Bars of the hoof (E---)
- Frog (E---)
- Sole (EB--)

Structures of the claw (--CF)
- (Note the "quick" on an unpigmented claw)

Injection landmarks for
- Bursas
- Carpal and digital muscle tendon sheaths
- Joint capsules
- Nerve blocks

RADIOGRAPHS

Scapula
- Spine (--CF)
  - Acromion
    - Hamate process
    - Suprahamate process (---F)
  - Supraglenoid tubercle
  - Neck
  - Glenoid cavity

Clavicle (---F)
- Dog: occasionally present

Humerus
- Head
- Major tubercle
- Minor tubercle
- Intermediate tubercle (E---)
- Crest of the major tubercle
  - Deltoid tuberosity
- Medial epicondyle
- Lateral epicondyle
- Supratrochlear foramen (--C-)
- Supracondylar foramen (---F)
- Trochlea
- Nutrient foramen

Ulna
- Tuber olecrani
- Anconeal process
  - Dog: a separate ossification center in some large breeds
- Trochlear notch
- Coronoid process
- Radial notch (--CF)
- Styloid process (lateral)
  - Horse: may normally appear separated from radius by small line

Radius
- Head
- Styloid process

Carpus
- Carpal bones
  - Accessory
  - Ulnar
  - Radial (EB--)
    - Horse: appears distal to intermediate carpal bone on a flexed lateral view
  - Intermedioradial (--CF)
  - Intermediate (EB--)
    - Horse: appears proximal to radial carpal bone on a flexed lateral view
  - First (--CF)
  - Second (E-CF)
  - Second and third (-B--)
  - Third (E-CF)
  - Fourth
  - Fifth (E---)
    - Horse: bone is present in a small number of animals
  - Horse: note the amount of movement possible at each joint on a flexed lateral view

Metacarpus
- Metacarpal bones
  - First (--CF)
  - Second (E-CF)
  - Third (E-CF)
  - Third and fourth (-B--)
  - Fourth (E-CF)
  - Fifth (-BCF)
  - Horse: note which oblique view shows each bone best

Digits
- Phalanges
  - Proximal
  - Middle
  - Distal
    - Canal of the sole (E---)
    - Extensor process (EB--)
    - Marginal notch (E---)
- Sesamoid bones
  - Dorsal (--CF)
  - Proximal (palmar)
  - Distal (palmar) (EB--)
- Calcar metacarpeum (E-CF)
  - Ergot (E---)
  - Metacarpal pad (--CF)
- Torus digitalis (-BCF)
  - Digital pads (--CF)

Phalanges, torus digitalis (cont.)
        Torus ungulae (dewclaw) (-B--)
            Bones within
    Frog (E---)
    Hoof wall (EB--)
        Horse: note normal relation with distal phalanx

2

PELVIC LIMB

OSTEOLOGY

Sacroiliac joint

Ossa coxarum (pelvis)
    Acetabulum
        Acetabular fossa
        Acetabular notch
        Lunate surface
    Ischiatic arch
    Ischiatic spine
    Obturator foramen
    Pelvic symphysis
        Ventral ridge (-B---)
    Ilium
        Tuber coxae
            Cranial ventral iliac spine
            Caudal ventral iliac spine
        Tuber sacrale
            Cranial dorsal iliac spine
            Caudal dorsal iliac spine
        Iliac crest
        Arcuate line
        Major (greater) ischiatic (sciatic) notch
    Ischium
        Minor (lesser) ischiatic (sciatic) notch
        Ischiatic tuberosity
    Pubis
        Iliopubic eminence
        Pubic tubercle
    Acetabular bone (--CF-)

Coxal (hip) joint

Os femoris (femur)
    Head
        Fovea capitis
    Neck
    Major (greater) trochanter
        Cranial part (E-c-P)
        Caudal part (E-c-P)

Os femoris, major trochanter (cont.)
        Trochanteric fossa
    Minor (lesser) trochanter
    Intertrochanteric line
    Third trochanter (E-C--)
    Supracondylar fossa (EB---)
    Medial and lateral supracondylar tuberosities (EBC-P)
    Medial and lateral epicondyles
    Medial condyle
    Lateral condyle
        Extensor fossa
        Fossa for the popliteal m.
    Intercondylar fossa
    Trochlea

Genual (femorotibial & femoropatellar) (stifle) joint

Tibia
    Medial and lateral condyles
    Intercondylar eminences
        Medial and lateral intercondylar tubercles
    Popliteal notch
    Extensor groove
    Tibial tuberosity
    Cranial border
    Medial malleolus
    Lateral malleolus (E----)

Fibula
    Lateral malleolus
        Fused with tibia (E----)
        Separate small bone (-B---)

Sesamoid bones
    Patella
    2 in gastrocnemius m. (--CF-)
    1 in popliteus m. (--CF-)

Tarsocrural joint

Tarsal bones
    Talus
        Trochlea
    Calcaneus
        Tuber calcanei
        Sustentaculum tali
    Central
        Fused with fourth (-B---)
    First
        Fused with second (E----)

Second
Fused with first (E----)
Fused with third (-B---)
Third
Fused with second (-B---)
Fourth
Fused with central (-B---)
Tarsal canal

(Proximal intertarsal joint)

(Distal intertarsal joint)

Tarsometatarsal joints

Metatarsal bones
First (--CF-)
Second
Third
Fused with fourth (-B---)
Fourth
Fused with third (-B---)
Fifth (--CFP)
Metatarsal sesamoid bone of third interosseous m. (-B---)

Metatarsophalangeal joint

Proximal (plantar) sesamoid bones

Dorsal sesamoid bones (--CF-)

Phalanges
Proximal phalanx
Proximal interphalangeal joint
Middle phalanx
Distal interphalangeal joint
Distal (plantar) sesamoid bone (EB--P)
Distal phalanx
Axial foramen (-B---)
Extensor process (EB--P)
Medial and lateral plantar processes (EB--P)
Parietal grooves (E----)
Marginal notch (crena) (E----)
Semilunar line (E----)
Solar foramen to canal of the sole (E----)
Base (--CF-)
Ungual crest (--CF-)
Ungual process (--CF-)
Phalanges of the dewclaws (-B--P)

Pelvic muscles
- Tensor fasciae latae m.
  - Origin
    - Tuber coxae
  - Termination
    - Fascia lata
    - Lateral patellar lig. (E----)
  - Function
    - Flex coxal joint
  - Innervation
    - Cranial gluteal n.
  - Blood supply
    - Lateral circumflex femoral a.
    - Iliolumbar a.
    - Deep circumflex iliac a.
    - Superficial circumflex iliac a. (--C--)

- Superficial gluteal m. (E-CFP)
  - Origin
    - Tuber coxae (E----)
    - Sacral vertebrae (--CFP)
    - Caudal vertebrae (--CF-)
  - Termination
    - Third trochanter (E-C--)
    - Major trochanter (---FP)
  - Function
    - Abduct limb
    - Extend coxal joint (--CF-)
    - Flex coxal joint (E----)
  - Innervation
    - Caudal gluteal n.
  - Blood supply
    - Caudal gluteal a.
  - Bursa
    - Trochanteric bursa is located between tendon and major trochanter (E-cfp)

- Middle gluteal m.
  - Origin
    - Wing of ilium
    - Longissimus dorsi m. (EB---)
    - Broad sacrotuberous lig. (EB---)
    - Sacral and caudal vertebrae (---F-)
  - Termination
    - Major trochanter
  - Function
    - Abduct limb
    - Extend coxal joint

Innervation
Cranial gluteal n.
Ischiatic n. (-B---)
Blood supply
Iliolumbar a. (EBCF-)
Cranial gluteal a.
Bursa
Trochanteric bursa is located between tendon and major trochanter (EB--P)
Miscellaneous
Accessory head

Deep gluteal m.
Origin
Body of ilium
Ischiatic spine
Tuber coxae (-B---)
Broad sacrotuberous lig. (-B---)
Termination
Major trochanter
Function
Abduct limb
Extend coxal joint (--CFP)
Innervation
Cranial gluteal n.
Blood supply
Cranial gluteal a.

Internal obturator m. (E-CF-)
Origin
Dorsal surface of the ischium and pubis
Termination
Trochanteric fossa
Function
Rotate limb outward
Innervation
Ischiatic n.
Blood supply
Obturator a.
Bursa
Between tendon and minor ischiatic notch
Miscellaneous
Tendon crosses minor ischiatic notch

External obturator m.
Origin
Ventral surface of ischium and pubis
Dorsal surface of ischium and pubis (intrapelvic part) (-B--P)
Termination
Trochanteric fossa

Pelvic muscles, external obturator m. (cont.)

Function
Rotate limb outward
Innervation
Obturator n.
Blood supply
Obturator a.

Quadratus femoris m.
Origin
Ventral ischium
Termination
Intertrochanteric line
Function
Extend coxal joint
Rotate limb outward
Innervation
Ischiatic n.
Blood supply
Medial circumflex femoral a.
Caudal gluteal a. (--CF-)

Gemelli mm.
Origin
Minor ischiatic notch
Termination
Trochanteric fossa
Function
Rotate limb outward
Innervation
Ischiatic n.
Blood supply
Caudal gluteal a. (-BCFP)
Obturator a. (E----)

Piriformis m. (--cFp)
Origin
Sacral and caudal vertebrae
Sacrotuberous lig. (--cF-)
Broad sacrotuberous lig. (----p)
Termination
Third trochanter (--c--)
Distal to major trochanter (---Fp)
Function
Extend coxal joint
Innervation
Caudal gluteal n.
Blood supply
Caudal gluteal a.
Cranial gluteal a. (--CF-)

Thigh muscles
- Iliopsoas m.
  - Origin
    - Psoas major m.
    - Iliacus m.
  - Termination
    - Minor trochanter
  - Function
    - Flex coxal joint
  - Innervation
    - Lumbar nn.
  - Blood supply
    - Lumbar aa.
    - Medial circumflex femoral a.
    - Lateral circumflex femoral a.
    - Iliolumbar a.

- Biceps femoris m.
  - Origin
    - Ischiatic tuberosity
    - Sacroiliac lig. (E----)
    - Broad sacrotuberous lig. (-B--P)
    - Sacrotuberous lig. (--C--)
    - Sacral vertebrae (-B---)
    - Caudal vertebrae (---F-)
  - Termination
    - Common calcanean tendon
    - Cranial border of tibia
    - Fascia lata
    - Lateral patellar lig. (E----)
  - Function
    - Extend coxal joint
    - Extend genual joint
    - Flex genual joint
    - Extend tarsus
  - Innervation
    - Ischiatic n.
    - Caudal gluteal n.
  - Blood supply
    - Caudal gluteal a.
    - Caudal femoral a.
      - Distal caudal femoral a. (--CF-)
  - Bursa
    - Located between muscle and major trochanter (Eb---)
  - Miscellaneous
    - Gluteobiceps m. (-B---)
      - Fused biceps femoris and superficial gluteal mm.

- Semitendinosus m.
  - Origin
    - Ischiatic tuberosity

Thigh muscles, semitendinosus m., origin (cont.)
- Caudal vertebrae (E---P)
- Termination
  - Common calcanean tendon
  - Cranial border of tibia (EB--P)
  - Medial tibia (--CF-)
- Function
  - Extend coxal joint
  - Flex genual joint
  - Extend tarsus
- Innervation
  - Tibial n.
- Blood supply
  - Caudal gluteal a. (E-CFP)
  - Caudal femoral a. (E-CFP)
    - Distal caudal femoral a. (--CF-)
  - Medial circumflex femoral a. (-B--P)

Semimembranosus m.
- Origin
  - Ischiatic tuberosity
  - Broad sacrotuberous lig. (E----)
- Termination
  - Craniomedial os femoris (-BCFP)
  - Medial epicondyle of os femoris (E----)
  - Medial condyle of tibia (-BCFP)
- Function
  - Extend coxal joint
  - Flex genual joint (-BCFP)
- Innervation
  - Tibial n.
- Blood supply
  - Medial circumflex femoral a.
  - Caudal gluteal a. (E-CFP)
- Miscellaneous
  - 2 distinct bellies (-BCFP)

Sartorius m.
- Origin
  - Wing of ilium (-BCFP)
  - Psoas minor tendon (E---P)
- Termination
  - Medial femoral fascia
  - Medial patellar lig. (E----)
- Function
  - Flex coxal joint
- Innervation
  - Femoral n.
- Blood supply
  - Femoral a. (E---P)
  - Saphenous a. (-BCF-)
  - Superficial circumflex iliac a. (--C--)

Miscellaneous
- 2 separate parts (--C--)
- Femoral vessels pass through the muscle (-B--P)

Gracilis m.
- Origin
  - Symphyseal (subpelvic) tendon
- Termination
  - Medial femoral fascia
  - Common calcanean tendon (--CF-)
  - Medial patellar lig. (E----)
- Function
  - Adduct limb
  - Extend tarsus (--CF-)
- Innervation
  - Obturator n.
- Blood supply
  - Medial circumflex femoral a. (EB--P)
  - Femoral a. (-BCF-)
    - Proximal caudal femoral a. (--CF-)
- Miscellaneous
  - Partially fused with adductor m. (---P)

Pectineus m.
- Origin
  - Iliopubic eminence
- Termination
  - Distomedial os femoris (EBC-P)
  - Medial epicondyle of os femoris (-B---)
  - Distal to minor trochanter (---F-)
- Function
  - Adduct limb
- Innervation
  - Obturator n.
- Blood supply
  - Medial circumflex femoral a.
- Miscellaneous
  - 2 separate bellies (-B---)

Adductor m.
- Origin
  - Symphyseal (subpelvic) tendon
  - Cranial pubis (---F-)
- Termination
  - Midcaudal os femoris
  - Medial patellar lig. (E----)
- Function
  - Adduct limb
- Innervation
  - Obturator n.

Thigh muscles, adductor m. (cont.)

- Blood supply
  - Medial circumflex femoral a.
  - Caudal femoral a.
    - Distal caudal femoral a. (--CF-)
- Miscellaneous
  - 2 separate bellies (ebCF-)
  - Partially fused with gracilis m. (----P)

- Quadriceps femoris m.
  - Rectus femoris m.
    - Origin
      - Body of ilium
    - Termination
      - Patella
        - Patellar tendon to tibial tuberosity (--CFP)
        - Medial, lateral, and intermediate patellar ligg. to proximal tibia (EB---)
    - Function
      - Flex coxal joint
      - Extend genual joint
    - Innervation
      - Femoral n.
    - Blood supply
      - Lateral circumflex femoral a.
      - Superficial circumflex iliac a. (--C--)
  - Vastus lateralis, medialis, intermedius mm.
    - Origin
      - Proximal os femoris
    - Termination
      - Patella
        - Patellar tendon to tibial tuberosity (--CFP)
        - Medial, lateral, and intermediate patellar ligg. to proximal tibia (EB---)
    - Function
      - Extend genual joint
    - Innervation
      - Femoral n.
    - Blood supply
      - Lateral circumflex femoral a.

Crus muscles

- Cranial tibial m.
  - Origin
    - Craniolateral tibia
  - Termination
    - Proximoplantar metatarsals (-BC--)
    - Proximal metatarsals (---FP)
    - Third metatarsal (E----)
    - Medial first tarsal (Cunean tendon) (E----)
      - Bursa under tendon
    - Second tarsal (----P)
      - Bursa under tendon

Function
- Flex tarsus

Innervation
- Deep peroneal n.

Blood supply
- Cranial tibial a.

Miscellaneous
- Distal tendon passes through peroneus tertius tendon (EB--)

Long digital extensor m.

Origin
- Extensor fossa

Termination
- Distal phalanx
- Middle phalanx of third digit (EB--P)
- Proximal phalanx of third digit (E----)

Function
- Extend digits
- Flex tarsus

Innervation
- Deep peroneal n.

Blood supply
- Cranial tibial a.

Lateral digital extensor m.

Origin
- Proximal tibia (EBC--)
- Proximal fibula

Termination
- Long digital extensor tendon (--CFP)
- Middle phalanx of lateral digit (-B---)
- Proximal phalanx (E-cf-)

Function
- Extend digits
- Flex tarsus

Innervation
- Superficial peroneal n.

Blood supply
- Cranial tibial a.

Peroneus longus m. (-BCFP)

Origin
- Proximal tibia
- Proximal fibula (-BCF-)

Termination
- Proximoplantar metatarsals (-BCF-)
- Fourth tarsal bone (----P)

Function
- Flex tarsus

Innervation
- Deep peroneal n.

Crus muscles, peroneus longus m. (cont.)
- Blood supply
  - Cranial tibial a.

Peroneus tertius m. (EB--P)
- Origin
  - Extensor fossa
- Termination
  - Proximal metatarsals (EB--p)
  - First tarsal bone (----P)
  - Second tarsal bone (----P)
- Function
  - Flex tarsus when genual joint is flexed (E----)
  - Flex tarsus (-B--P)
  - Extend genual joint (-B--P)
- Innervation
  - Deep peroneal n.
- Blood supply
  - Cranial tibial a.
- Miscellaneous
  - Distal attachment is split for cranial tibial tendon to pass through (EB---)
  - Tendinous throughout (E----)

Peroneus brevis m. (--CF-)
- Origin
  - Distal tibia
  - Distal fibula
- Termination
  - Proximal metatarsal V
- Function
  - Flex tarsus
- Innervation
  - Superficial peroneal n.
- Blood supply
  - Cranial tibial a.

Gastrocnemius m.
- Origin
  - Medial and lateral supracondylar tuberosities (EBC-P)
  - Medial and lateral condyles of os femoris (---F-)
- Termination
  - Tuber calcanei with common calcanean tendon
- Function
  - Flex genual joint
  - Extend tarsus
- Innervation
  - Tibial n.
- Blood supply
  - Caudal femoral a.
    - Distal caudal femoral a. (--CF-)
- Miscellaneous
  - Sesamoid in each head of origin (--CF-)

- Tendon twists to lateral side of superficial digital flexor tendon

Superficial digital flexor m.
- Origin
  - Supracondylar fossa (EB---)
  - Lateral supracondylar tuberosity (--C-P)
  - Lateral condyle of os femoris (---F-)
- Termination
  - Tuber calcanei with common calcanean tendon
  - Middle phalanx
  - Proximal phalanx (E----)
- Function
  - Flex genual joint (-BCFP)
  - Extend tarsus (-BCFP)
  - Extend tarsus when genual joint is extended (E----)
  - Flex digits (EBC-P)
- Innervation
  - Tibial n.
- Blood supply
  - Caudal femoral a.
    - Distal caudal femoral a. (--CF-)
- Bursa
  - Superficial bursa between skin and tendon at the tarsus
  - Calcaneal bursa between tendon and tuber calcanei
- Miscellaneous
  - Tendinous throughout (E----)
- Feline modifications
  - Short digital flexor m.
    - Origin
      - Tendon of superficial digital flexor m. distal to tuber calcanei
    - Termination
      - Middle phalanx
    - Function
      - Flex digits
    - Innervation
      - Tibial n.

Deep digital flexor m.
- Origin
  - Proximal tibia
  - Proximal fibula
- Termination
  - Distal phalanx
    - Semilunar line (E----)
- Function
  - Flex digits
- Innervation
  - Tibial n.
- Blood supply
  - All heads
    - Cranial tibial a. (-BCFP)

Crus muscles, deep digital flexor m., blood supply, all heads (cont.)
- Caudal tibial a. (E----)
- Flexor digiti I longus m.
  - Caudal tibial a.
- Miscellaneous
  - Flexor digiti I (hallucis) longus m. (lateral head)
    - Crosses sustentaculum tali
  - Long digital flexor m. (medial or deep head)
    - Joins lateral head at distal tarsus
  - Caudal tibial m. (superficial head) (EBefP)
  - Accessory (check) lig. attaches below tarsus (E----)

Popliteus m.
- Origin
  - Fossa for the popliteus m.
- Termination
  - Proximocaudal tibia (EBC-P)
  - Proximomedial tibia (---F-)
- Function
  - Flex genual joint
- Innervation
  - Tibial n.
- Blood supply
  - Caudal femoral a. (E-CFP)
    - Distal caudal femoral a. (--CF-)
  - Caudal tibial a. (EB---)
  - Cranial tibial a. (-B---)
- Miscellaneous
  - Sesamoid in tendon of origin (--CF-)

Soleus m. (eB-FP)
- Origin
  - Proximal fibula
  - Lateral epicondyle
- Termination
  - Gastrocnemius tendon
- Function
  - Assist gastrocnemius m.
- Innervation
  - Tibial n.
- Blood supply
  - Caudal femoral a.
    - Distal caudal femoral a. (---F-)

Tarsal muscles

Interflexorius m. (--CF-)
- Origin
  - Deep digital flexor m. distal to tarsus
- Termination
  - Short digital flexor tendons
- Function
  - Flex digits

Short digital extensor m.
Origin
Proximal metatarsals (EBCF-)
Termination
Long digital extensor tendons
Function
Extend digits
Innervation
Deep peroneal n.

LIGAMENTS

Sacroiliac ligg. (EB--) (Fig. 29)
Pelvic attachment
Tuber sacrale
Vertebral attachment
Sacrum
Miscellaneous
Form lateral border of pelvic canal
Have ventral, dorsal and interosseus parts

Broad sacrotuberous lig. (EB--) (Fig. 29)
Pelvic attachment
Ischiatic tuberosity
Ischiatic spine
Tuber coxae
Vertebral attachment
Sacrum
Caudal vertebrae

Sacrotuberous lig. (--C-) (Fig. 29)
Pelvic attachment
Ischiatic tuberosity
Vertebral attachment
Sacrum
Caudal vertebrae

Lig. of the femoral head (EbCF)
Femoral attachment
Fovea capitis
Pelvic attachment
Acetabular fossa

Accessory lig. (E---)
Proximal attachment
Cranial pubic lig.
Distal attachment
Fovea capitis
Miscellaneous
Passes through acetabular notch

Transverse acetabular lig.
    Crosses acetabular notch to complete acetabular lip

Patellar lig.
    Proximal attachment
        Patella
    Distal attachment
        Tibial tuberosity
    Miscellaneous
        Medial, intermediate and lateral patellar ligg. (EB--)

Femoropatellar ligg.
    Caudal attachment
        Medial and lateral femoral epicondyles
    Cranial attachment
        Patella

Collateral ligg. of the genual joint
    Proximal attachment
        Medial and lateral femoral epicondyles
    Distal attachment
        Proximal, medial and lateral tibia
        Proximal fibula

Cranial cruciate lig.
    Proximal attachment
        Caudomedial part of lateral femoral condyle
    Distal attachment
        Craniomedial tibial intercondyloid fossa
    Function
        Prevent tibia from sliding craniad

Caudal cruciate lig.
    Proximal attachment
        Craniolateral part of medial femoral condyle
    Distal attachment
        Caudolateral tibia
    Function
        Prevent tibia from sliding caudad

Proximal extensor retinaculum of the tarsus

Middle extensor retinaculum of the tarsus (E---)

Distal extensor retinaculum of the tarsus

Lateral extensor retinaculum of the tarsus

Flexor retinaculum of the tarsus
    Flexor canal

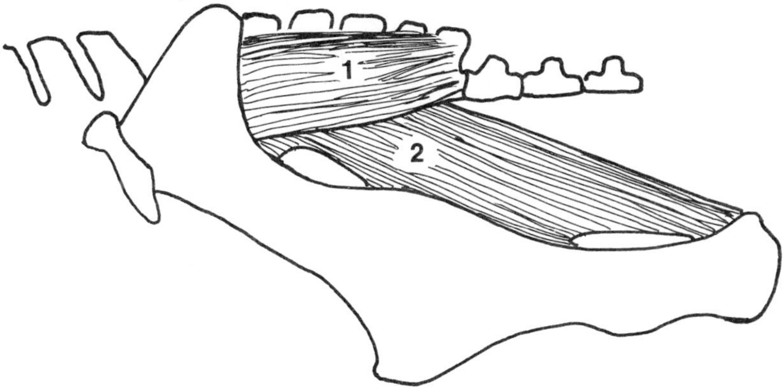

Horses

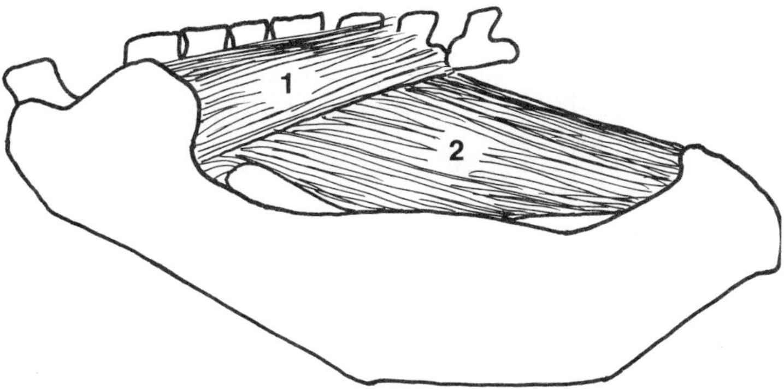

Cattle

Dogs

Figure 29. Pelvic ligaments

1. Sacroiliac ligament
2. Broad sacrotuberous ligament
3. Sacrotuberous ligament

Medial and lateral menisci
- Associated ligaments
  - Transverse lig. (--CF)
  - Cranial and caudal tibial ligg. for each meniscus
  - Meniscofemoral lig. of the lateral meniscus

Plantar tarsal ligg.
- Proximal attachment
  - Tuber calcanei
- Distal attachment
  - Proximal metatarsals

Medial and lateral collateral ligg. of the tarsus
- Proximal attachment
  - Tibia
  - Fibula (-BCF)
- Distal attachment
  - Proximal metatarsals

Dorsal lig. (E---)
- Distomedial tarsus

Accessory lig. (E---)
- Distal continuation of tarsal joint capsule
- Joins deep digital flexor tendon

SYNOVIAL SHEATHS

Genual synovial tendon sheaths
- Long digital extensor tendon sheath
  - Extension of medial femorotibial joint capsule

Tarsal synovial tendon sheaths (Fig. 30)
- 6 of the 7 (or 8) muscles crossing the tarsus are usually associated with a tendon sheath
  - Cranial tibial tendon sheath
    - Shared with
      - Peroneus tertius m. (EB--)
    - Medial (Cunean) tendon has a bursa under its distal attachment (EB--)
  - Long digital extensor tendon sheath
  - Lateral digital extensor tendon sheath
    - Shared with
      - Peroneus brevis m. (--CF)
  - Peroneus tertius tendon sheath
    - Shared with
      - Cranial tibial m. (EB--)
  - Peroneus longus tendon sheath (-BCF)
    - Shared with
      - Cranial tibial m. (--CF)

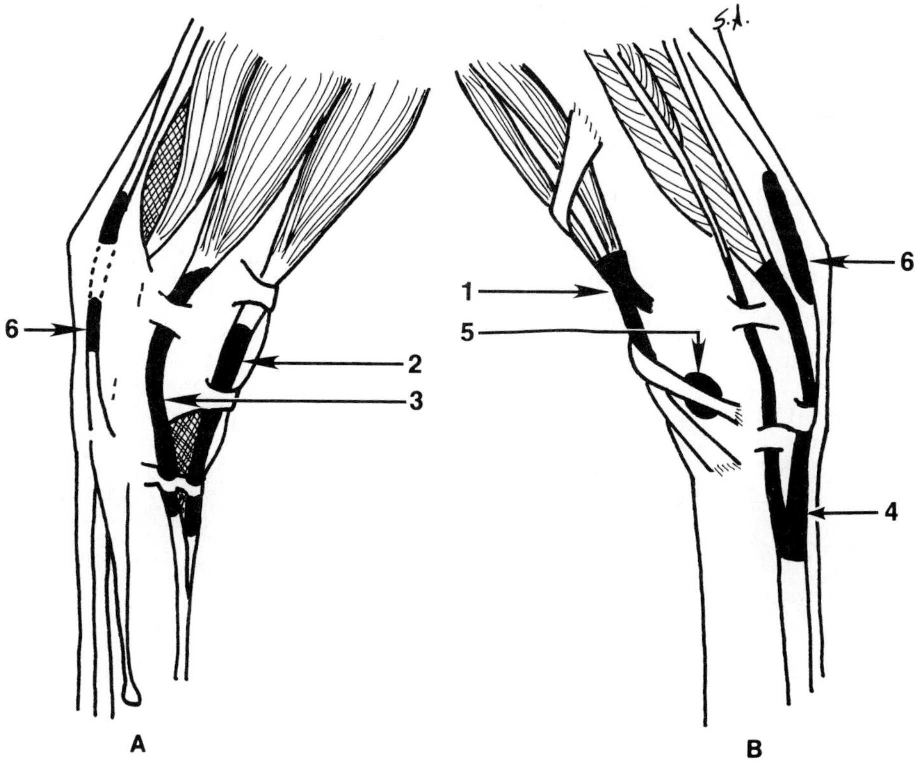

Figure 30. Synovial structures of the equine tarsus (A-lateral aspect, B-medial aspect)

1. Cranial tibial and peroneus tertius tendon sheath
2. Long digital extensor tendon sheath
3. Lateral digital extensor tendon sheath
4. Deep digital flexor tendon sheath
5. Bursa of the medial (Cunean) tendon
6. Bursa of the superficial digital flexor tendon

Tarsal synovial tendon sheaths, muscles crossing tarsus (cont.)
- Peroneus brevis tendon sheath (--CF)
  - Shared with
    - Lateral digital extensor m.
- Superficial digital flexor m. (no tendon sheath)
- Deep digital flexor tendon sheath

Proximal positions
- Usually begin just proximal to tarsus

Distal extent
- Usually end just distal to tarsus

Digital synovial tendon sheath
- Surrounds the superficial and deep digital flexor tendons

JOINT CAPSULES

Coxal joint capsule
- Injection sites
  - Notch of the major trochanter is landmark for approach (E---)
  - Insert needle cranial to dorsal aspect of the major trochanter, directing it medially and perpendicular to median plane

Genual joint capsules
- Femoropatellar capsule
  - Communicates with medial femorotibial sac in adults
  - May communicate with lateral femorotibial sac (does in dogs)
- Femorotibial capsule
  - Medial femorotibial sac
    - Equally divided by medial meniscus
    - Communicates with femoropatellar capsule
    - Communicates with lateral femorotibial sac (-bCF)
    - Injection site
      - Insert needle just cranial to medial collateral lig.
  - Lateral femorotibial sac
    - Equally divided by lateral meniscus
    - May communicate with femoropatellar capsule (does in dogs and cats)
    - Communicates with medial femorotibial sac (-bCF)
    - Injection site
      - Insert needle just cranial to lateral collateral lig.

Tarsal joint capsules
- Tarsocrural capsule
  - Communicates with
    - Talocalcaneal capsule
    - (Proximal intertarsal capsule)
      - Talocalcaneocentral capsule

- Calcaneoquartal capsule
- Capsular sacs
  - Dorsomedial sac
    - Bounded by
      - Medial malleolus
      - Tendon of peroneus tertius m.
      - Medial tendon of cranial tibial m. (E---)
      - Long part of medial collateral lig.
    - Crossed by cranial branch of medial saphenous v.
    - Injection site
      - Insert needle behind the vein; direct it laterally (E---)
      - Insert needle in mediolateral direction (-B--)
  - Medioplantar sac
    - Bounded by
      - Long part of medial collateral lig.
      - Tendon of deep digital flexor m.
      - At level of the medial malleolus
  - Lateroplantar sac
    - Bounded by
      - Calcaneus
      - Lateral malleolus
- Talocalcaneal capsule
  - Includes joint formed between talus and calcanean bones
  - Communicates with
    - Tarsocrural capsule
    - Talocalcaneocentral capsule
    - Calcaneoquartal capsule
- (Proximal intertarsal capsule)
  - Talocalcaneocentral capsule
    - Includes joints between talus, central and calcanean bones
    - Communicates with
      - Tarsocrural capsule
      - Talocalcaneal capsule
      - Calcaneoquartal capsule
  - Calcaneoquartal capsule
    - Includes joints between calcaneus and fourth tarsal bones
    - Communicates with
      - Tarsocrural capsule
      - Talocalcaneal capsule
      - Talocalcaneocentral capsule
- Centrodistal (distal intertarsal) capsule
  - Includes joints between central tarsal bone and distal tarsal bones
- Tarsometatarsal capsule
  - Includes joints between proximal metatarsus and distal tarsal bones

ARTERIES

- External iliac a.
  - Deep circumflex iliac a. (EB--)
  - Uterine a. (E---)
  - Deep femoral a.
    - Pudendoepigastric trunk
      - Caudal epigastric a.
        - Supplies
          - Deep abdominal muscles
      - External pudendal a.
        - Supplies
          - Ventral genitalia
        - Caudal superficial epigastric a.
    - Medial circumflex femoral a.
      - Supplies
        - Medial thigh
      - Obturator a. (-BC-)

- Femoral a. (Figs. 31-34)
  - Lateral circumflex femoral a.
    - Supplies
      - Craniolateral thigh
  - Proximal caudal femoral a. (--CF)
    - Supplies
      - Medial thigh
  - Descending genicular a.
  - Saphenous a.
    - Cranial branch (E-CF)
      - Dorsal common digital aa. (--CF)
        - Dorsal proper digital aa.
    - Caudal branch
      - Medial plantar a.
        - Deep plantar arch
        - Plantar common digital aa. (-BCF)
          - Plantar proper digital aa.
        - Plantar common digital a. II (E---)
      - Lateral plantar a.
        - Deep plantar arch
          - Plantar metatarsal aa.
        - Plantar common digital a. III (E---)
        - Plantar common digital a. IV (-B--)
          - Abaxial plantar proper digital a. IV (-B--)
          - Axial plantar proper digital a. V (-B--)
  - Middle caudal femoral a. (--CF)
  - Distal caudal femoral a. (--CF)
    - Supplies
      - Popliteal region
  - Caudal femoral a. (EB--)
    - Supplies
      - Popliteal region

Popliteal a.
- Location
  - Caudal to genual joint
- Caudal tibial a.

Cranial tibial a.
- Superficial branch
  - Dorsal common digital aa. (-B--)
    - Dorsal proper digital aa.
- Dorsal pedal a.
  - Perforating tarsal a. (E---)
    - Deep plantar arch
      - Plantar metatarsal aa.
  - Perforating metatarsal a. (--CF)
    - Deep plantar arch
      - Plantar metatarsal aa.
  - Arcuate a. (--CF)
    - Dorsal metatarsal aa.
  - Dorsal metatarsal a. (EB--)
    - Plantar proper digital aa. (E---)
      - Terminal arch
    - Proximal perforating metatarsal a. (-B--)
      - Proximal deep plantar arch
    - Distal perforating metatarsal a. (-B--)
      - Distal deep plantar arch

(Note: Superficial vessels and nerves are called common digital in the region of the metatarsus and proper digital in the region of the phalanges. Deep vessels and nerves are called metatarsals.)

VEINS

Deep major veins follow the arteries (Figs. 35-38)

Lateral saphenous v. (eBCf)
- Joins the distal caudal femoral v. (--C-)
- Joins the femoral v. (eB-f)
  - Between the biceps femoris m. and the semitendinosus m.
- Branches
  - Cranial branch (--Cf)
  - Caudal branch (--Cf)

Medial saphenous v. (EbCF)
- Joins the femoral v.
- Branches
  - Cranial branch (E-CF)
  - Caudal branch (EbCF)

1. External iliac a.
2. Deep femoral a.
3. Pudendoepigastric trunk
4. Caudal epigastric a.
5. External pudendal a.
6. Caudal superficial epigastric a.
7. Medial circumflex femoral a.
9. Lateral circumflex femoral a.
10. Saphenous a.
13. Medial (plantar common) digital a.
14. Lateral (plantar common) digital a.
15. Descending genicular a.
16. Caudal femoral a.
19. Caudal tibial a.
20. Cranial tibial a.
22. Dorsal pedal a.
23. Dorsal metatarsal a.
24. Perforating tarsal a.
25. Medial and lateral (plantar) metatarsal aa.
27. Medial and lateral (plantar proper) digital aa.

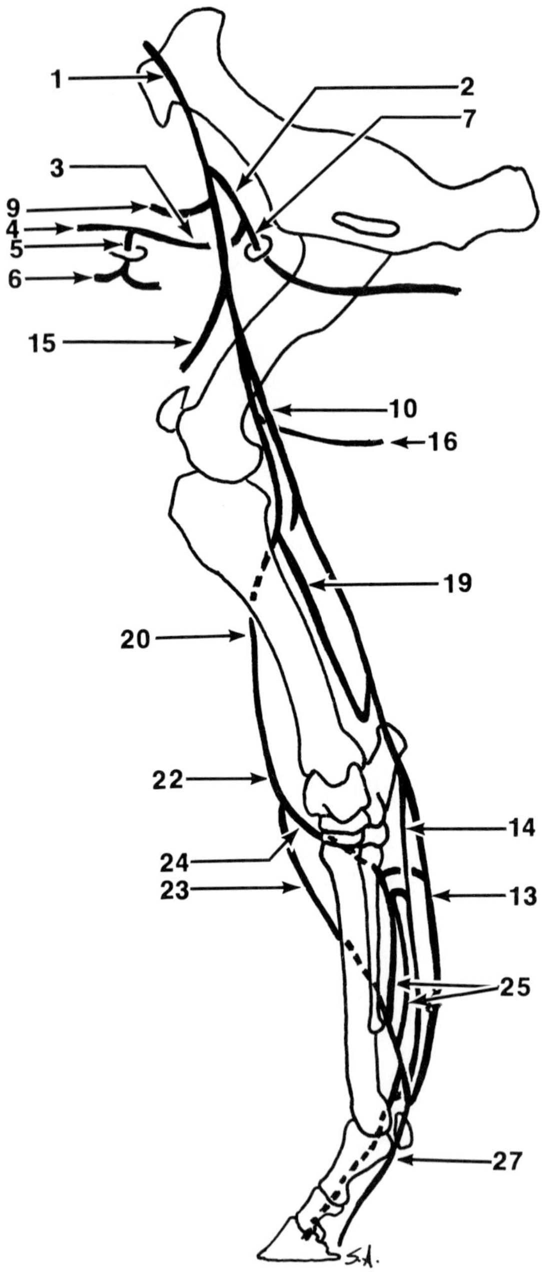

Figure 31. Pelvic limb arteries of the horse (medial aspect)

1. External iliac a.
2. Deep femoral a.
3. Pudendoepigastric trunk
4. Caudal epigastric a.
5. External pudendal a.
6. Caudal superficial epigastric a.
7. Medial circumflex femoral a.
9. Lateral circumflex femoral a.
10. Saphenous a.
13. Medial plantar common digital a.
14. Lateral plantar common digital a.
15. Descending genicular a.
16. Caudal femoral a.
20. Cranial tibial a.
22. Dorsal pedal a.
23. Dorsal metatarsal a.
25. Medial and lateral plantar metatarsal aa.
27. Plantar proper digital aa.
29. Dorsal proper digital aa.

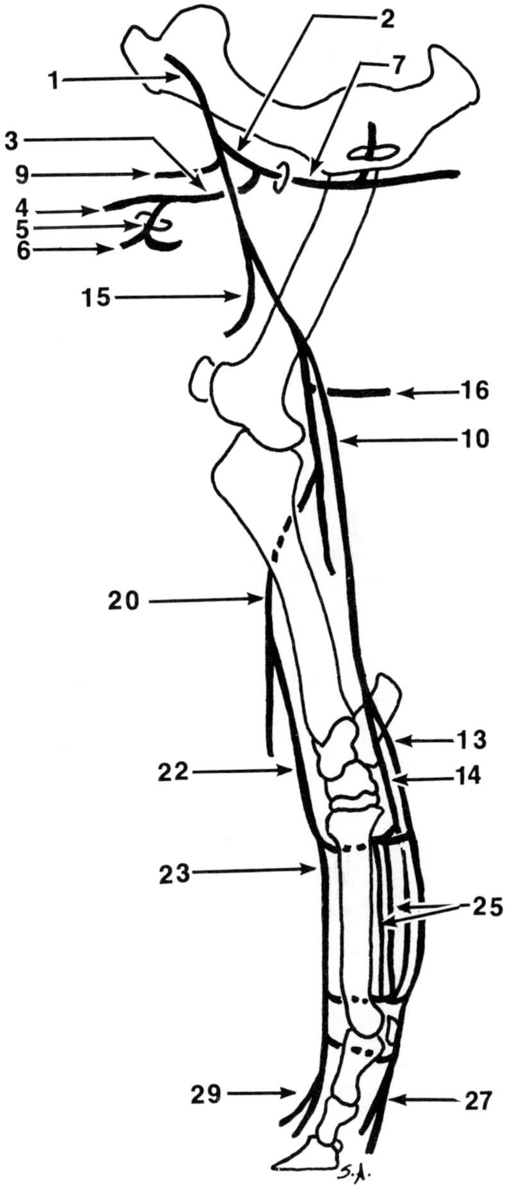

Figure 32. Pelvic limb arteries of cattle (medial aspect)

1. External iliac a.
2. Deep femoral a.
3. Pudendoepigastric trunk
4. Caudal epigastric a.
5. External pudendal a.
6. Caudal superficial epigastric a.
7. Medial circumflex femoral a.
9. Lateral circumflex femoral a.
10. Saphenous a.
11. Cranial branch
12. Caudal branch
15. Descending genicular a.
17. Distal caudal femoral a.
19. Caudal tibial a.
20. Cranial tibial a.
21. Superficial branch
22. Dorsal pedal a.
23. Dorsal metatarsal aa.
24. Perforating metatarsal a.
25. Plantar metatarsal aa.
26. Dorsal proper digital aa.
27. Plantar proper digital aa.
28. Dorsal common digital aa.
29. Dorsal proper digital aa.

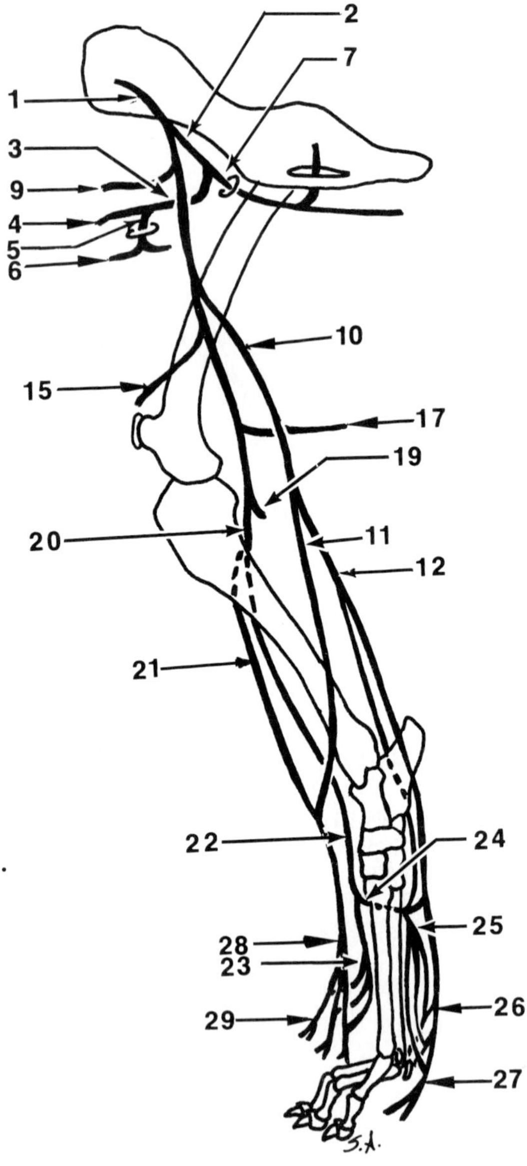

Figure 33. Pelvic limb arteries of the dog (medial aspect)

1. External iliac a.
2. Deep femoral a.
3. Pudendoepigastric trunk
4. Caudal epigastric a.
5. External pudendal a.
6. Caudal superficial epigastric a.
7. Medial circumflex femoral a.
9. Lateral circumflex femoral a.
10. Saphenous a.
11. Cranial branch
12. Caudal branch
15. Descending genicular a.
17. Distal caudal femoral a.
19. Caudal tibial a.
20. Cranial tibial a.
21. Superficial branch
22. Dorsal pedal a.
23. Dorsal metatarsal aa.
24. Perforating metatarsal a.
25. Plantar metatarsal aa.
26. Dorsal proper digital aa.
27. Plantar proper digital aa.
28. Dorsal common digital aa.
29. Dorsal proper digital aa.

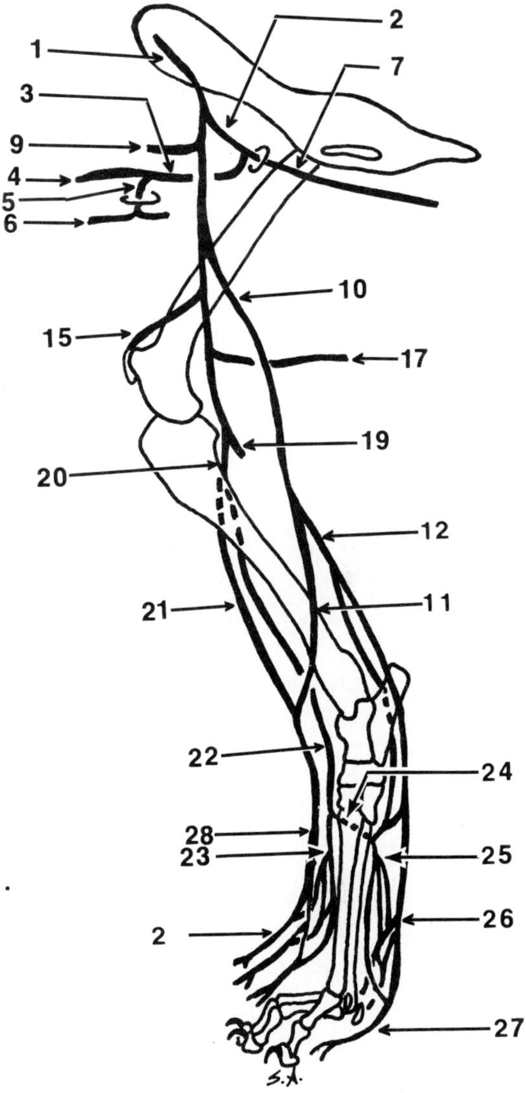

Figure 34. Pelvic limb arteries of the cat (medial aspect)

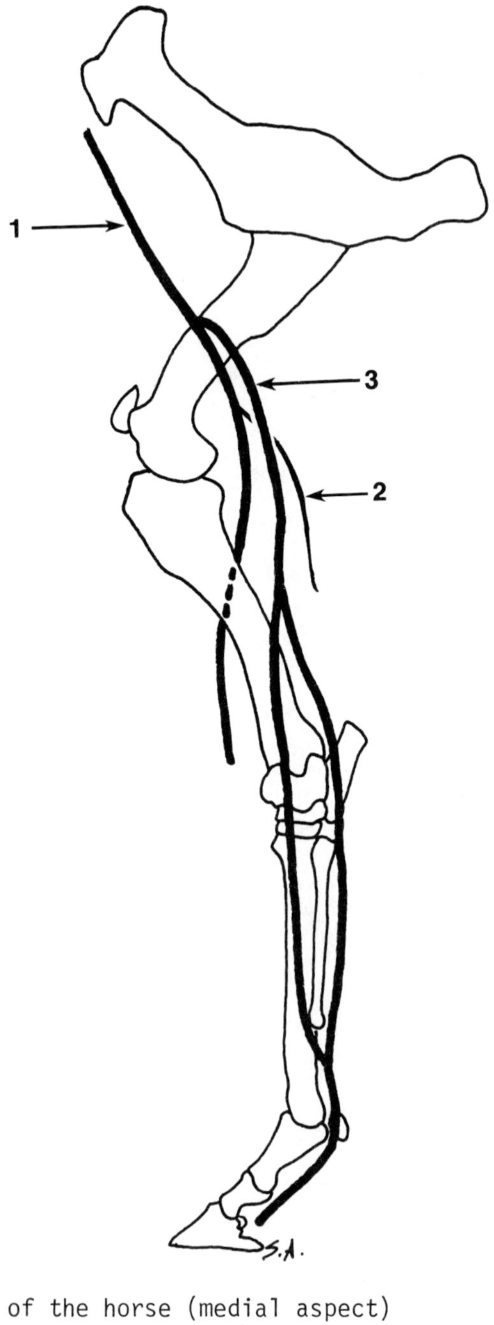

1. Femoral v.
2. Lateral saphenous v.
3. Medial saphenous v.

Figure 35. Pelvic limb veins of the horse (medial aspect)

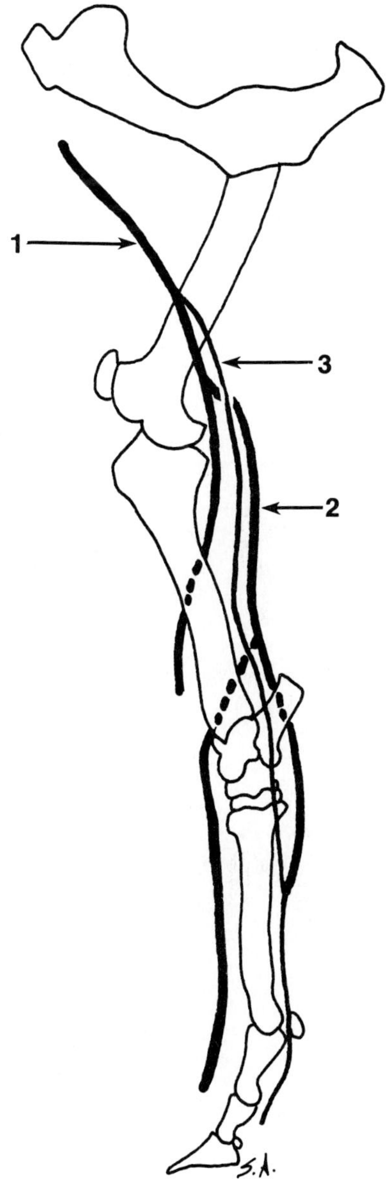

1. Femoral v.
2. Lateral saphenous v.
3. Medial saphenous v.

Figure 36. Pelvic limb veins of cattle (medial aspect)

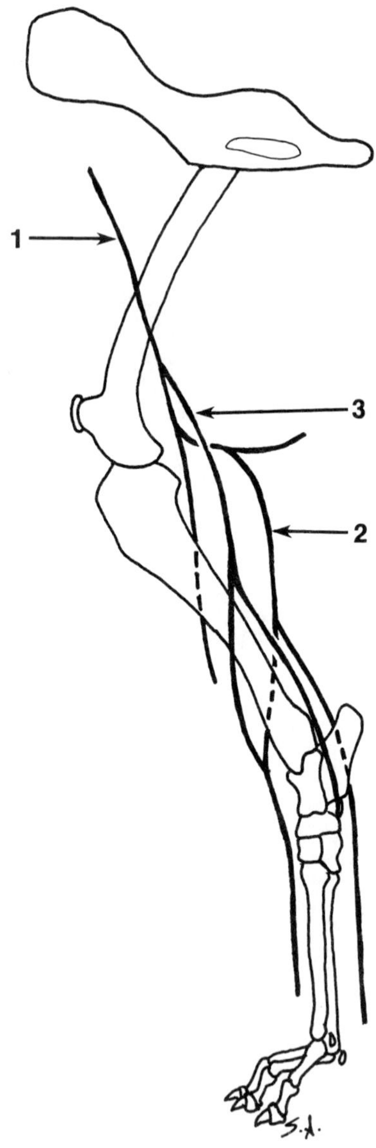

1. Femoral v.
2. Lateral saphenous v.
3. Medial saphenous v.

Figure 37. Pelvic limb veins of the dog (medial aspect)

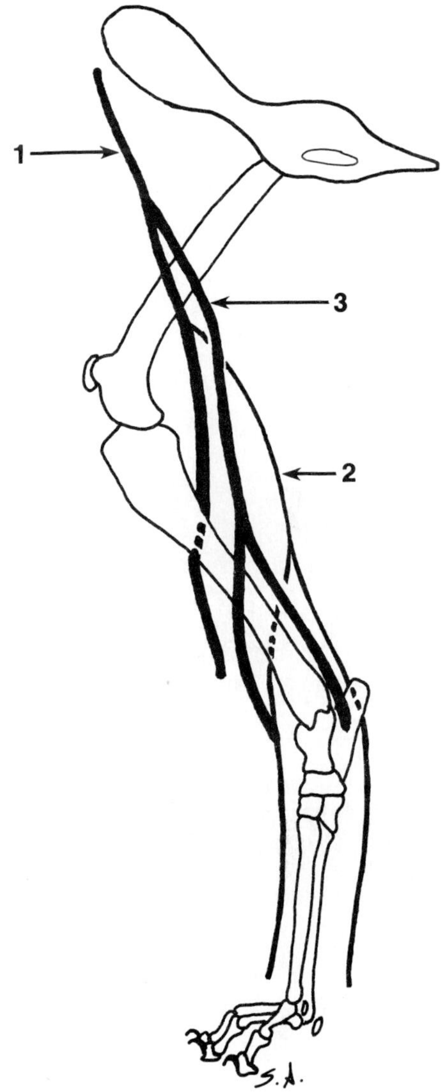

1. Femoral v.
2. Lateral saphenous v.
3. Medial saphenous v.

Figure 38. Pelvic limb veins of the cat (medial aspect)

NERVES

Cranial gluteal n.
Spinal nerve contributions
L_5 (e---)
L_6
L_7 (--CF)
S_1 (EBC-)
Supplies motor fibers to
Gluteal muscles
Sensory to
Lateral hip
Courses
Emerges over major ischiatic notch

Caudal gluteal n.
Spinal nerve contributions
L_6 (---F)
L_7 (--CF)
S_1 (EBcF)
S_2 (EBc-)
S_3 (eb--)
Supplies motor fibers to
Caudal thigh muscles
Sensory to
Caudolateral hip
Course
Lateral to sacrotuberous lig. (--C-)
Lateral to broad sacrotuberous lig. (EB--)

Femoral n.
Spinal nerve contributions
L_3 (e-c-)
L_4 (EBC-)
L_5
L_6 (eB-F)
Supplies motor fibers to
Quadriceps femoris m.
Sensory to (Figs. 39,40)
Craniomedial limb
Course of travel
Deep, near femoral triangle
Continues as the saphenous n. (Figs. 41-48)

Obturator n.
    Spinal nerve contributions
        L_3 (e---)
        L_4 (Ebc-)
        L_5 (EBC-)
        L_6 (ebCF)
        L_7 (---F)
    Supplies motor fibers to
        Adductors of limb
    Course of travel
        Crosses medial side of wing of the ilium
        Passes through cranial part of obturator foramen

Ischiatic (sciatic) n.
    Spinal nerve contributions
        L_5 (e---)
        L_6
        L_7 (--CF)
        S_1
        S_2 (eBcf)
    Supplies motor fibers to
        Distal muscles of limb
    Sensory to
        Cranial, lateral and caudal limb
    Course of travel (Figs. 41-48)
        Emerges over major ischiatic notch
        Travels medial to biceps femoris m., caudal to os femoris
    Branches
        Caudal cutaneous sural n.
            Sensory to
                Lateral tarsus and metatarsus
        Common peroneal n. (Figs. 39,40)
            Course of travel (Figs. 41-48)
                Crosses lateral head of gastrocnemius m.
            Injection site
                Between long and lateral digital extensor mm.
                    Approximately two-thirds of the way distal on leg region
                On craniolateral surface of gastrocnemius m.
                    Near origin of lateral head
            Branches
                Superficial peroneal n.
                    Sensory to
                        Cranial tarsus and metatarsus
                        Lateral tarsus and metatarsus
                        Medial tarsus and metatarsus (-B--)

Ischiatic n., branches, superficial peroneal n. (cont.)
            - Becomes the dorsal common digital nn.
                - Join the dorsal metatarsal nn. (--CF)
        - Deep peroneal n.
            - Supplies motor fibers to
                - Craniolateral crus muscles
            - Sensory to
                - Cranial metatarsus
                - Lateral metatarsus
            - Becomes the dorsal metatarsal nn.
                - Join dorsal common digital nn. (--CF)
                - Join plantar metatarsal nn. (EB--)
                    - Become the dorsal (proper) digital nn.
- Tibial n. (Figs. 39,40)
    - Supplies motor fibers to
        - Caudal muscles of crus
    - Course of travel (Figs. 41-48)
        - Passes between the heads of gastrocnemius m.
        - Lies between common calcanean tendon and deep digital flexor m.
    - Branches
        - Medial plantar n.
            - Sensory to
                - Medial metatarsus
                - Caudal metatarsus
            - Communicating branch to lateral plantar n. (E---)
            - Becomes the plantar common digital nn. (-BCF)
                - Join plantar metatarsal nn. (--CF)
                - Become the plantar (proper) digital nn.
            - Becomes the medial plantar common digital n. (E↓--)
        - Lateral plantar n.
            - Sensory to
                - Lateral metatarsus
                - Interosseous m. (E---)
            - Gives rise to the plantar metatarsal nn. (EB--)
                - Supplies interosseous m. (E---)
                - Join dorsal (proper) digital nn. (E---)
                    - Become the dorsal (proper) digital nn.
            - Receives communicating branch from medial plantar n. (E---)
            - Becomes the plantar metatarsal nn. (--CF)
                - Join plantar common digital nn. (--CF)
                - Become the plantar (proper) digital nn.

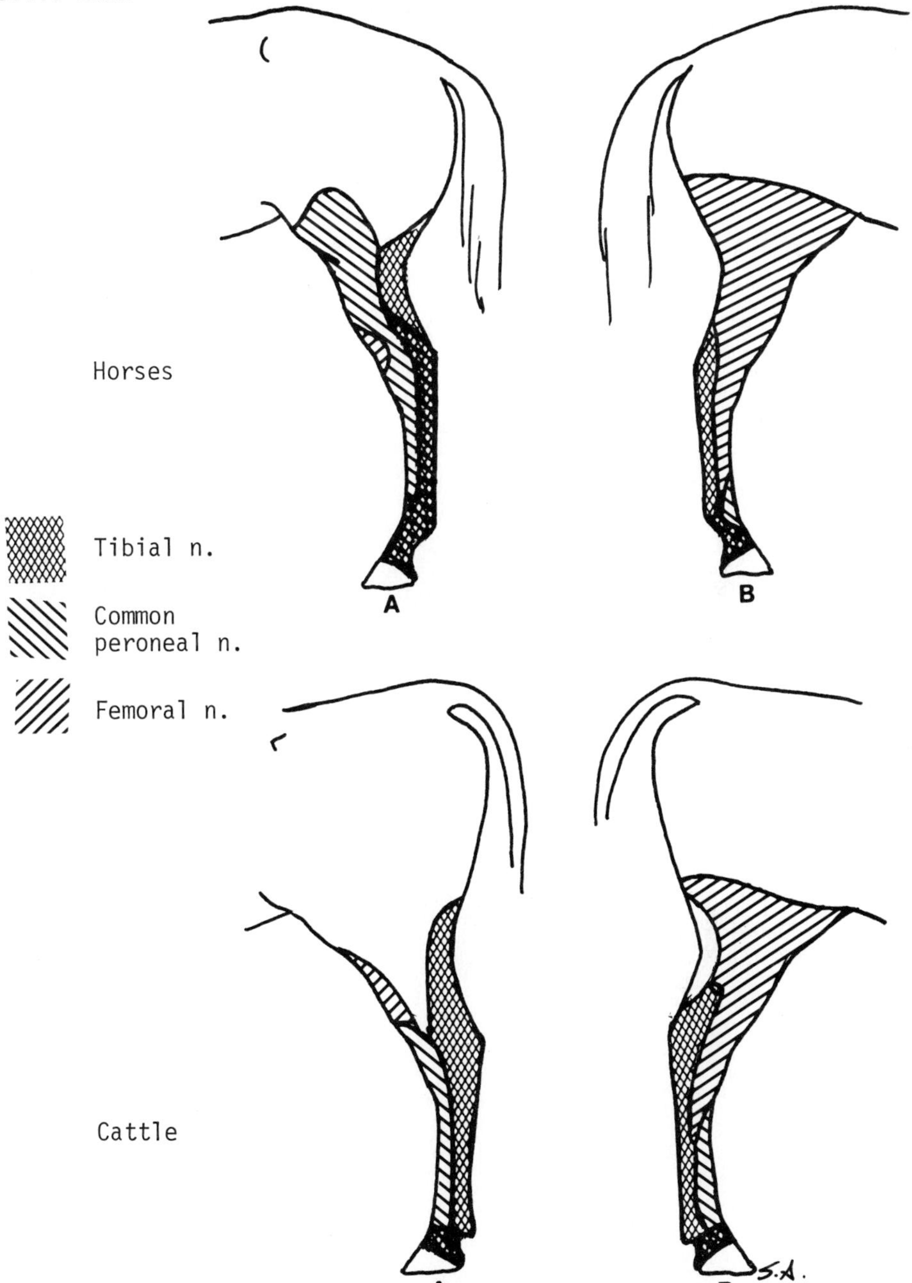

Figure 39. Cutaneous nerve supply of the equine and bovine pelvic limb (A-lateral aspect, B-medial aspect)

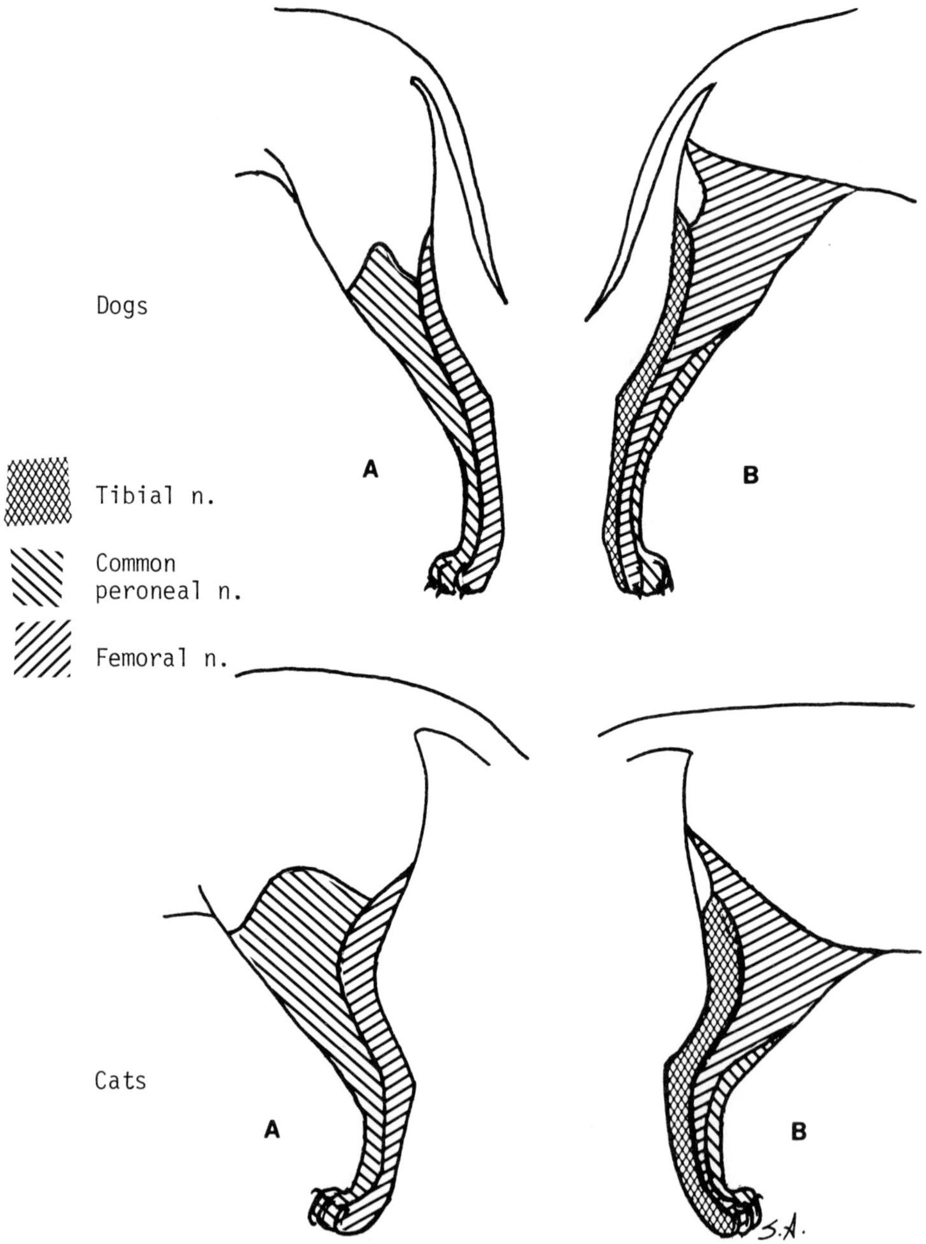

Figure 40. Cutaneous nerve supply of the canine and feline pelvic limb (A-lateral aspect, B-medial aspect)

1. Saphenous n.
2. Caudal cutaneous sural n. (of ischiatic n.)
3. Tibial n.
6. Medial plantar (common digital) n.
7. Lateral plantar (common digital) n.
9. Medial plantar metatarsal n.
10. Lateral plantar metatarsal n.
20. Communicating branch
21. Medial plantar (proper) digital n.
22. Dorsal branch
23. Lateral plantar (proper) digital n.
24. Dorsal branch

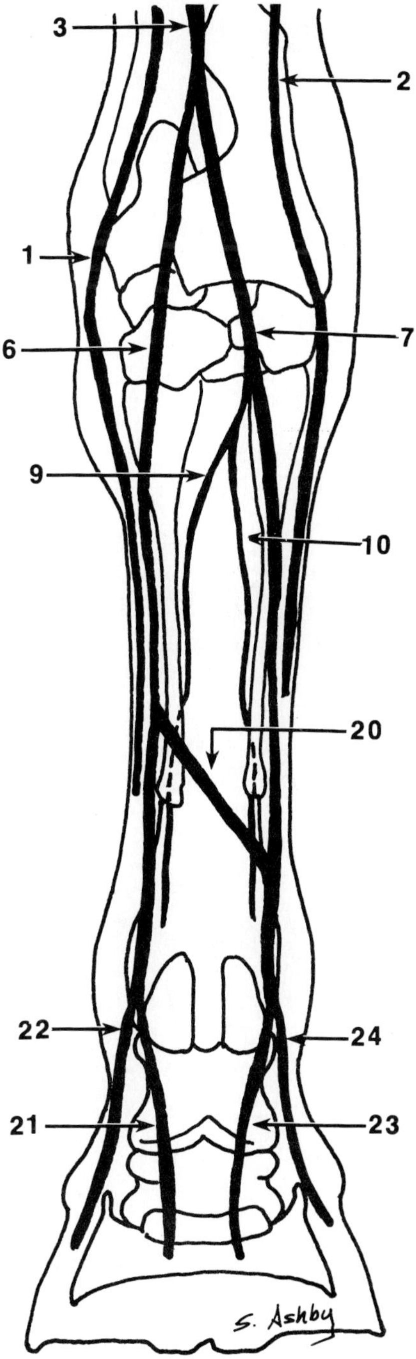

Figure 41. Nerves of the equine pes (right limb, plantar aspect)

2. Caudal cutaneous sural n. (od ischiatic n.)
3. Tibial n.
4. Superficial peroneal n.
5. Deep peroneal n.
7. Lateral plantar (common digital) n.
10. Lateral plantar metatarsal n.
12. Lateral dorsal (common digital) n.
19. Lateral dorsal metatarsal n.
20. Communicating branch
23. Lateral plantar (proper) digital n.
24. Dorsal branch
31. Lateral dorsal (proper) digital n.

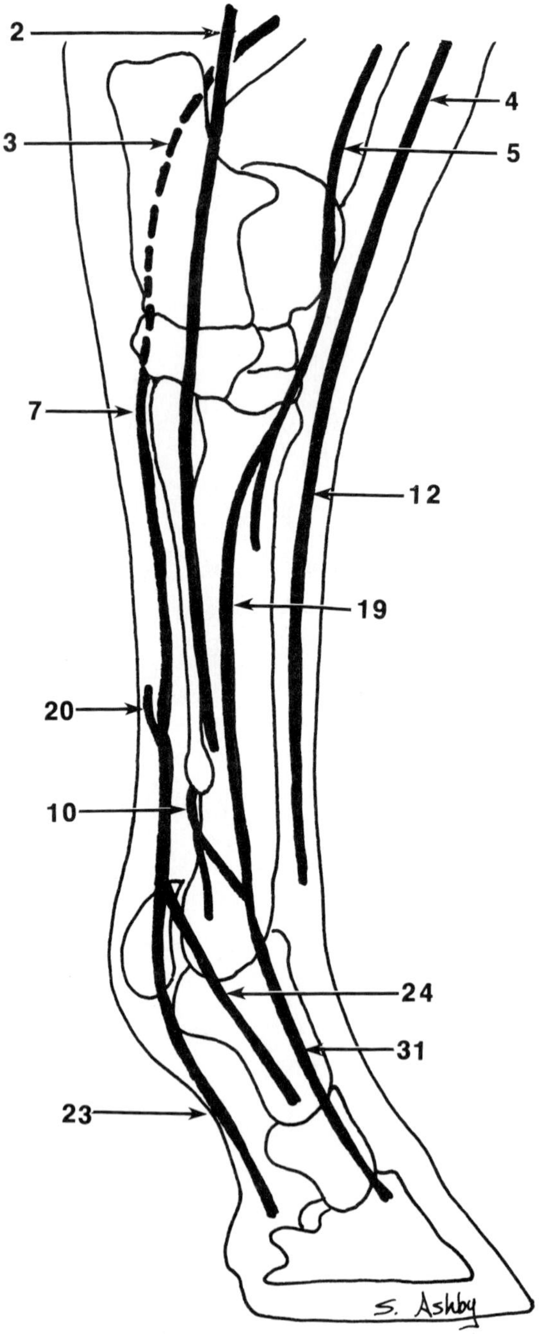

Figure 42. Nerves of the equine pes (right limb, lateral aspect)

1. Saphenous n.
3. Tibial n.
5. Deep peroneal n.
6. Medial plantar (common digital) n.
9. Medial plantar metatarsal n.
18. Medial dorsal metatarsal n.
20. Communicating branch
21. Medial plantar (proper) digital n.
22. Dorsal branch
30. Medial dorsal (proper) digital n.

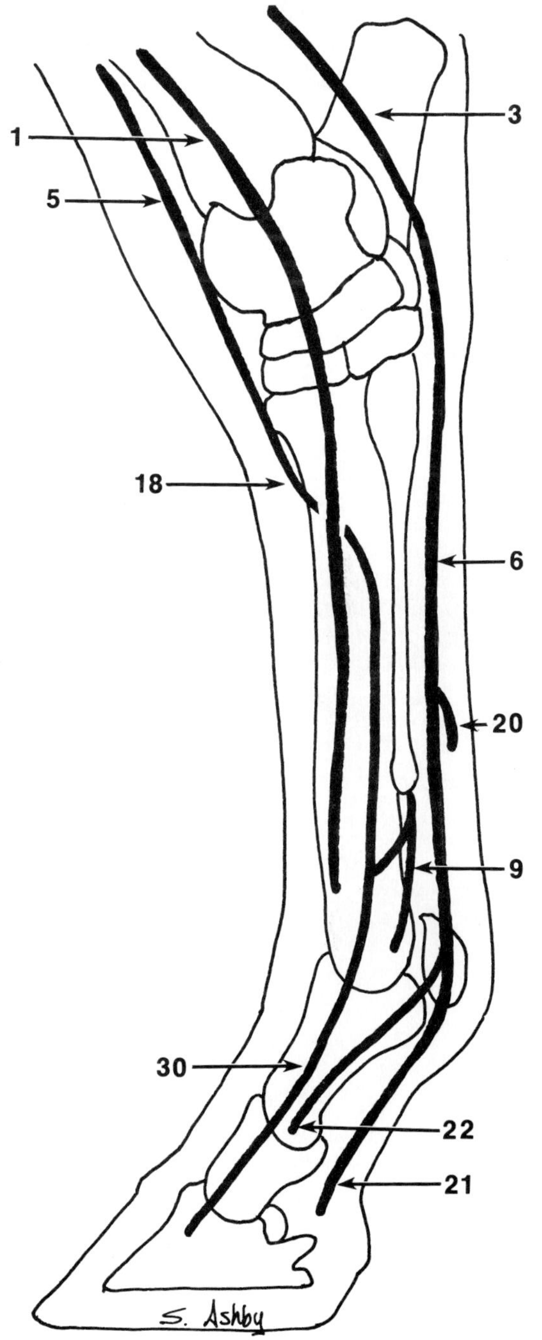

Figure 43. Nerves of the equine pes (right limb, medial aspect)

1. Saphenous n.
3. Tibial n.
6. Medial plantar common digital n.
7. Lateral plantar common digital n.
9. Medial plantar metatarsal n.
10. Lateral plantar metatarsal n.
25. Axial plantar proper digital n. III
26. Abaxial plantar proper digital n. III
27. Axial plantar proper digital n. IV
28. Abaxial plantar proper digital n. IV

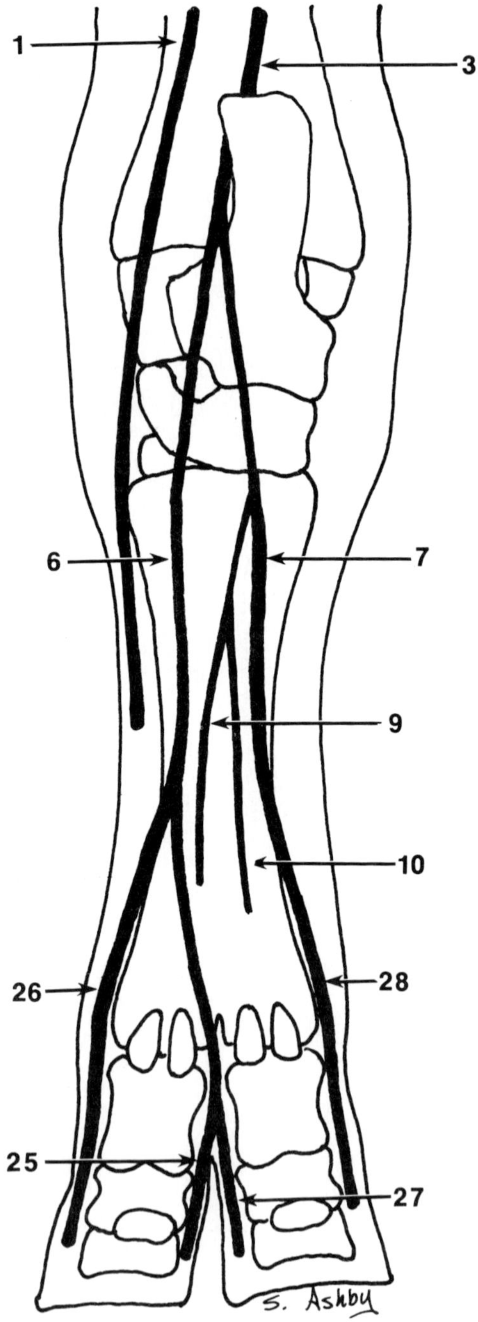

Figure 44. Nerves of the bovine pes (right limb, plantar aspect)

3. Tibial n.
4. Superficial peroneal n.
5. Deep peroneal n.
7. Lateral plantar common digital n.
13. Dorsal common digital n. IV
14. Dorsal common digital n. III
17. Dorsal metatarsal n.
28. Abaxial plantar proper digital n. IV
32. Axial dorsal proper digital n. III
34. Axial dorsal proper digital n. IV
35. Abaxial dorsal proper digital n. IV

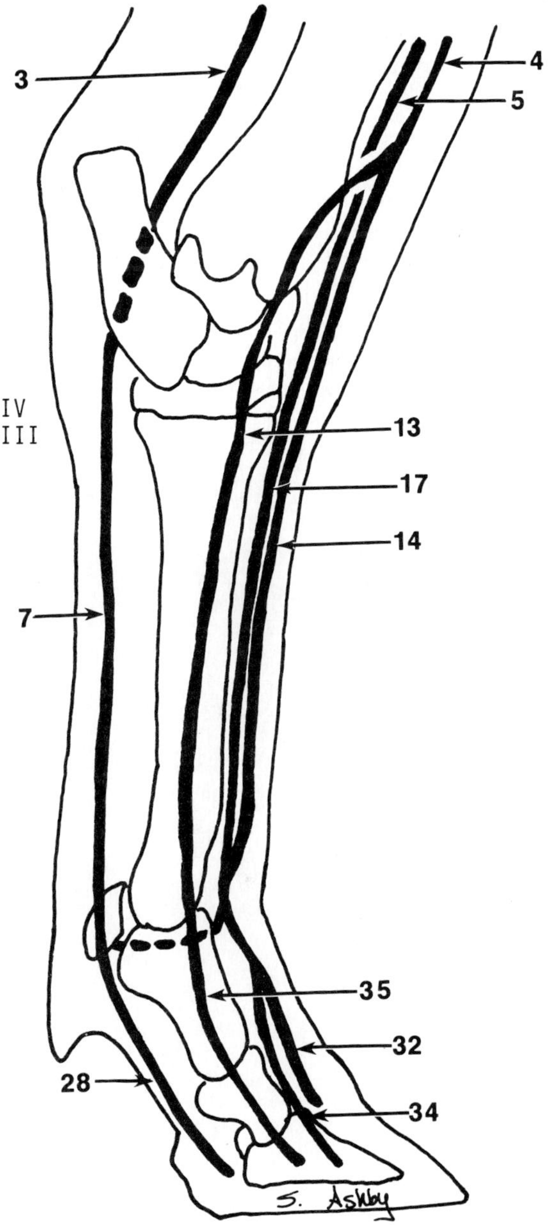

Figure 45. Nerves of the bovine pes (right limb, lateral aspect)

1. Saphenous n.
3. Tibial n.
4. Superficial peroneal n.
5. Deep peroneal n.
6. Medial plantar common digital n.
14. Dorsal common digital n. III
15. Dorsal common digital n. II
17. Dorsal metatarsal n.
25. Axial plantar proper digital n. III
26. Abaxial plantar proper digital n. III
27. Axial plantar proper digital n. IV
32. Axial dorsal proper digital n. III
33. Abaxial dorsal proper digital n. III
34. Axial dorsal proper digital n. IV

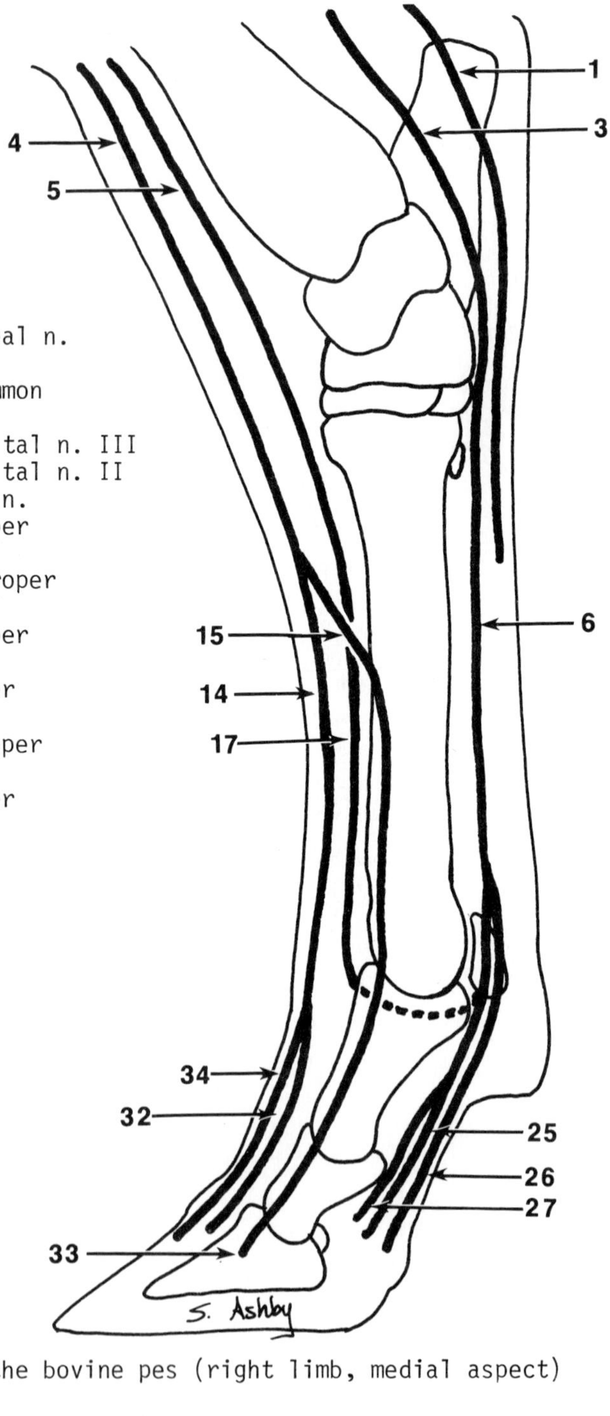

Figure 46. Nerves of the bovine pes (right limb, medial aspect)

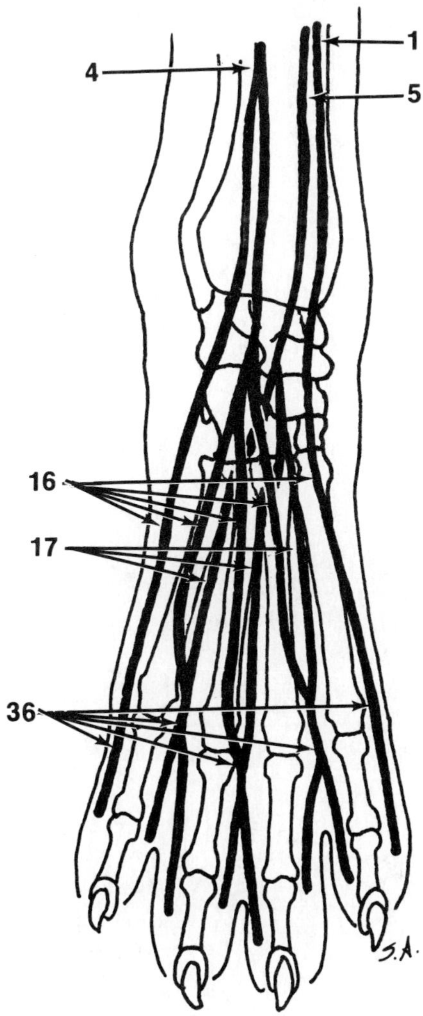

Figure 47. Nerves of the canine and feline pes (right limb, dorsal aspect)

1. Saphenous n.
4. Superficial peroneal n.
5. Deep peroneal n.
16. Dorsal common digital nn.
17. Dorsal metatarsal nn.
36. Dorsal proper digital nn.

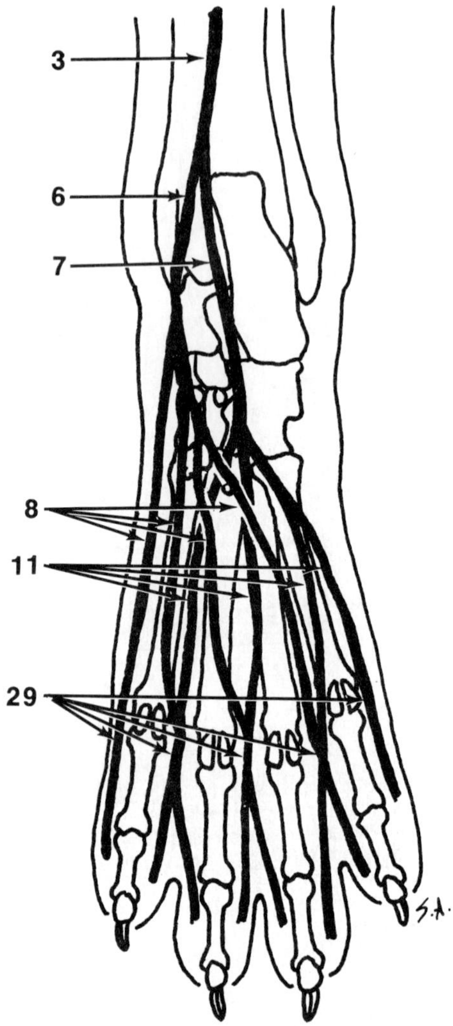

Figure 48. Nerves of the canine and feline pes (right limb, dorsal aspect)

3. Tibial n.
6. Medial plantar (common digital) n.
7. Lateral plantar (common digital) n.
8. Plantar common digital nn.
11. Plantar metatarsal nn.
29. Plantar proper digital nn.

LYMPH NODES

Popliteal ln.
- Location
  - Caudal to genual joint
- Area drained
  - Distal to genual joint
- Drains to
  - Iliofemoral ln. (E---)
  - Medial iliac ln. (-B-F)
  - Ischiatic ln. (-B--)
  - Inguinofemoral ln. (--CF)

Inguinofemoral (superficial inguinal) ln.
- Location
  - Near superficial inguinal ring
- Area drained
  - Caudal mammae
  - External genitalia
  - Popliteal ln. (--CF)
  - Medial thigh and crus (-B--)
- Drains to
  - Medial iliac ln. (-BCF)
  - Iliofemoral ln. (E---)

Iliofemoral (deep inguinal) ln. (E---)
- Location
  - Near deep inguinal ring
- Area drained
  - Pelvic limb
  - Abdominal mm.
  - Popliteal ln.
  - Inguinofemoral ln.
- Drains to
  - Medial iliac ln.

Ischiatic ln. (EB--)
- Location
  - Lateral to broad sacrotuberous lig.
- Area drained
  - Tail
  - Thigh
  - Anorectal ln. (E---)
  - Popliteal ln. (-B--)
- Drains to
  - Medial iliac ln.

Subiliac ln.
- Location
  - Cranial to thigh
- Area drained
  - Lumbar and gluteal regions
  - Pelvic limb

Subiliac ln. (cont.)
    Drains to
        Medial iliac ln.

Gluteal ln. (-B--)
    Location
        Proximomedial to biceps femoris m.
    Area drained
        Pelvic region
        Thigh
    Drains to
        Ischiatic ln.
        Medial iliac ln.

PALPATION

Ossa coxarum
    Tuber sacrale (-B--)
    Tuber coxae
    Ischiatic tuberosity
Os femoris
    Major trochanter
        Is in a line connecting tuber coxae and ischiatic tuberosity (E-CF)
        Is below a line connecting tuber coxae and ischiatic tuberosity (-B--)
    Medial epicondyle
    Trochlea
        Medial ridge (EB--)
Thigh muscles
    Tensor fasciae lata m. (E---)
    Quadriceps femoris m.
        Patellar ligg.
    Biceps femoris m.
    Semitendinosus m. (E---)
    Semimembranosus m. (E---)
    Gracilis m.
    Pectineus m.
Broad sacrotuberous lig. (-B--)
Sacrotuberous lig. (--C-)
Subiliac ln. (EB--)
Femoral a. (--CF)
Popliteal ln. (--CF)
Patella
Medial and lateral collateral ligg. of the genual jt.
Tibia and fibula
    Tibial tuberosity
    Cranial border of the tibia
    Medial malleolus
    Lateral malleolus

Muscles of the crus and their tendons
- Gastrocnemius m.
  - Twist in tendon
  - Medial sesamoid (--CF)
- Superficial digital flexor m.
- Deep digital flexor m.
- Lateral digital extensor m. (EB--)
- Cranial tibial m.
  - Medial tendon (E---)
- Long digital extensor m.

Medial saphenous v. (E--F)
Lateral saphenous v. (-BC-)
Tibial n.
- Between the deep digital flexor m. and common calcanean tendon

Tuber calcanei
Trochlea of the talus
Lateral collateral lig. of the tarsus
Plantar lig.
Tarsocrural joint capsule sacs
Dorsal metatarsal a. (E---)
Torus tarseus (E-CF)
- Chestnut (E---)
- Tarsal pad (--CF)
- (Note position relative to tarsus)

Calcar metatarseum (E-CF)
- Ergot (E---)
- Metatarsal pad (--CF)

Torus digitalis (-BCF)
- Torus ungulae (dewclaw) (-B--)
- Digital pads (--CF)

RADIOGRAPHS

Ossa coxarum (--CF)
- Sacroiliac jt.
- Acetabulum
- Obturator foramen
- Wing of the ilium
- Ischiatic tuberosity
- Ischiatic arch
- Pelvic symphysis

Os femoris
- Head
- Neck
- Major trochanter
- Minor trochanter
- Intertrochanteric line
- Trochlea
  - Medial ridge (EB--)
- Condyles

Genual joint
- (Note normal joint spaces)

Sesamoids
- Patella
  - (Note relationship to trochlea)
- Sesamoids of the gastrocnemius m. (--CF)
  - 1 proximal to each femoral condyle
    - Lateral sesamoid is more proximal
- Sesamoid of the popliteus m. (--CF)
  - Proximal to fibula

Crus region
- Tibia
  - Condyles
  - Intercondylar eminences
  - Tibial tuberosity
  - Cranial border
  - Medial malleolus
  - Lateral malleolus (E---)
  - (Note position of proximal tibia relative to os femoris)
- Fibula
  - Proximal portion
    - Horse: May contain a normal radiolucent, non-ossified line
  - Lateral malleolus (-BCF)
- Gastrocnemius m. (--CF)
- Common calcanean tendon
- Popliteal ln. (--CF)
- Patellar ligg.

Tarsus
- Tarsal bones
  - Talus
  - Calcaneus
  - Central (E-CF)
  - Central and fourth (-B--)
  - First (-BCF)
  - First and second (E---)
  - Second (-BCF)
  - Third (E-CF)
  - Fourth (E-CF)
  - Horse: note which angular view shows cranial surface of each bone best
- Tarsal canal
- Soft tissue
  - Torus tarseus (chestnut) (E---)
- Horse: note amount of movement possible at each joint on a flexed lateral view

Metatarsus
- Metatarsal bones
  - First (--CF)
  - Second
  - Third (E-CF)
  - Third and fourth (-B--)

Fourth (E-CF)
Fifth (--CF)
Horse: note which angular view shows each bone best
Metatarsal sesamoid bone (-B--)
Calcar metatarseum (metatarsal pad) (--CF)

3

HEAD AND NECK

SKULL

Occipital bone
- Condylar canal
- Condyles
- Foramen magnum
- Hypoglossal foramen
- Interparietal process (--CF-)
- Jugular process
- Nuchal crest (E-CFP)
- Nuchal line (-B---)
- Occipital protuberance
- Groove of transverse sinus

Sphenoid bone (presphenoid + basispenoid bones)
- Alar canal
- Body
- Caudal alar foramen
- Foramen lacerum
- Foramen orbitorotundum (-B--P)
- Optic canal
- Orbital fissure (E-CF-)
- Oval foramen
- Rostral alar foramen (E-CF-)
- Round foramen (E-CF-)
- Sella turcica
- Wings

Pterygoid bone

Temporal bone
- Petrosal part
  - Cerebellar fossa
  - Internal acoustic meatus
  - Mastoid process
  - Styloid process
  - Stylomastoid foramen
  - Tympanic canal
- Squamous part
  - Mandibular fossa
  - Retroarticular foramen
  - Retroarticular process
  - Zygomatic process

Temporal bone (cont.)
- Tympanic part
  - Carotid canal
  - External acoustic meatus
  - Musculotuberal canal
  - Tympanic bulla

Parietal bone

Frontal bone
- Cornual process
- Ethmoid foramen
- Supraorbital foramen
- Zygomatic process

Ethmoid bone
- Dorsal nasal concha (turbinate)
- Ethmoid labyrinth
- Middle nasal concha (turbinate) (--CF-)

Ventral nasal concha (turbinate)

Vomer bone

Nasal bone

Lacrimal bone
- Lacrimal bulla (-B---)
- Lacrimal foramen
- Fossa for ventral oblique m.

Maxilla
- Alveoli dentales
- Caudal palatine foramen
- Facial tuberosity (-B---)
- Infraorbital canal
- Infraorbital foramen
- Lacrimal canal
- Maxillary foramen
- Sphenopalatine foramen

Palatine bone
- Major palatine foramen
- Minor palatine foramen

Incisive bone
- Alveoli dentalis
- Palatine fissure

Os rostrale (----P)

- Zygomatic bone
  - Frontal process
  - Temporal process

- Mandible
  - Body
    - Alveoli dentales
    - Mental foramen
      - Lateral side
      - Medial side (----P)
  - Ramus
    - Condyloid process
      - Head
    - Coronoid process
    - Mandibular angle
      - Angular process
    - Mandibular foramen
    - Masseteric fossa
  - Symphysis

- Hyoid bones
  - Thyrohyoid
  - Basihoid
    - Lingual process (EB--p)
  - Ceratohoid
  - Epihoid
  - Stylohoid
  - Tympanohyoid

- General features
  - Facial crest (Eb--p)
  - Orbit
  - Sagittal crest
  - Temporal line
  - Tentorium ossium
  - Tympanooccipital fissure
  - Zygomatic arch

ORAL CAVITY AND PHARYNX

Teeth

	Deciduous formulas	Permanent formulas
Horses	$2 \times \frac{3\ 0\ 3}{3\ 0\ 3} = 24$	$2 \times \frac{3\ 1\ (3\text{-}4)\ 3}{3\ 1\ \ 3\ \ 3} = 40\text{-}42$
Cattle	$2 \times \frac{0\ 0\ 3}{4\ 0\ 3} = 20$	$2 \times \frac{0\ 0\ 3\ 3}{4\ 0\ 3\ 3} = 32$
Dogs	$2 \times \frac{3\ 1\ 3}{3\ 1\ 3} = 28$	$2 \times \frac{3\ 1\ 4\ 2}{3\ 1\ 4\ 3} = 42$

Teeth (cont.)

	Deciduous formulas	Permanent formulas
Cats	$2 \times \frac{3\ 1\ 3}{3\ 1\ 2} = 26$	$2 \times \frac{3\ 1\ 3\ 1}{3\ 1\ 2\ 1} = 30$
Swine	$2 \times \frac{3\ 1\ 3}{3\ 1\ 3} = 28$	$2 \times \frac{3\ 1\ 4\ 3}{3\ 1\ 4\ 3} = 44$

Approximate eruption dates (birth, days, weeks, months, years)

Tooth number		Horses	Cattle	Dogs	Cats	Swine
Decid. inc.	1	6d	b	1m	3w	1m
	2	6w	b	1m	1m	2m
	3	6m	2w	1m	1m	B
	4	--	2w	--	--	--
Perm. inc.	1	2½y	2y	5m	4m	1y
	2	3½y	2½y	5m	4m	1½y
	3	4½y	3y	5m	4m	9m
	4	--	4y	--	--	--
Perm. Canine		3½-5y	--	6m	6m	9m
Perm. Prem.	1	6m	2½y	6m	--	5m
	2	2½y	2y	6m	4m	1¼y
	3	3y	2y	6m	6m	1¼y
	4	4y	--	6m	6m	1¼y
Perm. Molar	1	1y	6m	4m	6m	5m
	2	2y	1-1½y	5m	--	1y
	3	3½y	2½y	7m	--	1½y

Sectorial teeth (--CF)
- First lower molar
  - Two roots
- Last upper premolar
  - Three roots
    - Two rostral roots
    - One caudal root
    - Roots diverge (--C-)

Tonsils
- Lingual tonsil
  - Present as distinct follicles (EB--)
  - Present as diffuse tissue (--CF)
  - Located near root of the tongue
- Palatine tonsil
  - Present as distinct tonsils
  - Located on or in lateral wall of the oropharynx
  - Tonsillar fossa (E-CF)
  - Tonsillar sinus (-B--)
  - Semilunar fold (--CF)
    - Medial wall of the tonsillar fossa
- Paraepiglottic tonsil
  - Present as distinct tonsils (---f)
  - Located lateral to base of the epiglottis

Pharyngeal tonsil
    Present as distinct tonsils
    Located near the choanae
Sublingual tonsil
    Located ventral to the tongue
Tonsil of the soft palate
    Present as distinct tonsils (E---)
    Present as distinct follicles (-b--)
    Present as diffuse tissue (-BCF)
    Located on ventral surface of the soft palate
Tubal tonsil (EB--)
    Present as distinct tonsils (-B--)
    Present as diffuse tissue (E---)
    Located near pharyngeal opening of the auditory tube

Muscles
    Cricopharyngeus m.
        Origin
            Cricoid cartilage
        Termination
            Dorsal laryngopharynx
            Fibers blend with esophagus
    Thyropharyngeus m.
        Origin
            Thyroid cartilage
        Termination
            Dorsal laryngopharynx
            Fibers blend with esophagus
    Hyopharyngeus m.
        Origin
            Thyrohyoid bone
            Ceratohyoid bone
        Termination
            Dorsal laryngopharynx
            Fibers blend with esophagus

Tongue
    Lingual frenulum
        Sublingual caruncle
    Lyssa (--CF)
        Fibroelastic cord (E---)
            Similar structure to lyssa
    Papillae
        Conical papillae
        Filiform papillae
        Foliate papillae
            Contain taste buds
        Fungiform papillae
            Contain taste buds
        Lenticular papillae (-B--)
            Probably represent flattened conical papillae
        Vallate papillae
            Contain taste buds

LARYNX

- Laryngeal cartilages
    - Arytenoid cartilages
        - Corniculate process
        - Cuneiform process (--C-)
        - Muscular process
        - Paired cartilages
    - Cricoid cartilage
    - Epiglottic cartilage
    - Thyroid cartilage
        - Caudal cornu
        - Rostral cornu
        - Caudal thyroid incisure (E-CF)
- Vocal fold
- Vestibular fold
- Laryngeal ventricle (E-C-)
- Ligaments
    - Cricothyroid lig.
    - Cricotracheal lig.
    - Thyroepiglottic lig.
- Muscles
    - Cricothyroideus m.
    - Cricoarytenoideus dorsalis m.
    - Cricoarytenoideus lateralis m.
    - Thyroarytenoideus m.
        - Vocalis m.
    - Hyoepiglottic m.
- Nerves
    - Recurrent laryngeal n.
        - Caudal laryngeal n.
            - Motor to
                - All muscles of larynx except cricothyroideus m.
            - Course of travel
                - Origin is vagus n. near the thorax
                - Left nerve turns around aortic arch
                - Right nerve turns around right subclavian a.
    - Cranial laryngeal n.
        - Motor to
            - Cricothyroideus m.
        - Sensory to
            - Internal larynx
        - Course of travel
            - Origin is vagus n. near skull

NASAL CAVITY AND PHARYNX

- Nasal cavity
    - Nares
    - Nasal diverticulum (E---)
    - Nasolacrimal duct opening

- Nasolacrimal canal (Ebcf)
- Infraorbital canal
- Conchae
  - Dorsal concha
  - Middle concha
  - Ventral concha (EBC-)
  - Ethmoid labyrinth
- Meatus
  - Dorsal nasal meatus
  - Middle nasal meatus
  - Ventral nasal meatus
  - Common nasal meatus

Paranasal sinuses

- Conchal sinus (EBcf)
  - Opens into
    - Nasal cavity
    - Frontal sinus (conchofrontal sinus) (E---)
- Frontal sinus
  - Opens into
    - Nasal cavity
    - Maxillary sinus (frontomaxillary aperture) (E---)
    - Conchal sinus (conchofrontal sinus) (E---)
    - Palatine sinus (-B---)
  - Components
    - Frontal sinus (undivided) (e--F)
    - Rostral frontal sinus (-BC-)
      - Medial rostral frontal sinus (-B--)
      - Lateral rostral frontal sinus (-B--)
      - Intermediate rostral frontal sinus (-b--)
      - Medial frontal sinus (--C-)
      - Lateral frontal sinus (--C-)
    - Caudal frontal sinus (-B--)
      - Cornual diverticulum
      - Nuchal diverticulum
      - Postorbital diverticulum
- Maxillary sinus (EB--)
  - Opens into
    - Nasal cavity
    - Frontal sinus (frontomaxillary aperture) (E---)
    - Sphenopalatine sinus (E---)
    - Palatine sinus (maxillopalatine aperture) (-b--)
  - Lacrimal bulla (-B--)
  - Roots of the molars (E---)
  - Components (E---)
    - Rostral maxillary sinus
    - Caudal maxillary sinus
- Palatine sinus (EB--)
  - Opens into
    - Maxillary sinus (maxillopalatine aperture) (-B--)
    - Sphenoid sinus (sphenopalatine sinus) (E---)
  - Located in palatine bone

- Paranasal sinuses (cont.)
  - Sphenoid sinus (Eb-F)
    - Opens into
      - Nasal cavity
      - Palatine sinus (sphenopalatine sinus) (E---)
    - Located in presphenoid bone
  - Lacrimal sinus (-B--)
    - Opens into
      - Nasal cavity
      - Maxillary sinus
    - Located in lacrimal bone
  - Maxillary recess of the nasal cavity (--CF)

- Nasal pharynx
  - Auditory tube
    - Ostium
    - Diverticulum (guttural pouch) (E---)
      - Divided by stylohyoid bone into medial and lateral parts
      - Structures near the diverticulum wall
        - External carotid a.
        - Facial n.
        - Glossopharyngeal n.
          - Medial to external carotid a.
        - Hypoglossal n.
          - Lateral to external carotid a.
        - Internal carotid a.
        - Vagus n.

SENSE ORGANS

- Eye
  - Conjunctiva
    - Bulbar
    - Fornix
    - Palpebral
      - Semilunar fold of the conjunctiva (third eyelid)
        - Cartilage
        - Superficial lymph tissue
        - Superficial gland
          - Surrounds the cartilage
      - Lacrimal caruncle

  - Palpebrae
    - Cilia
    - Obicularis oculi m.
      - Attaches at each commissure
    - Levator palpebrae superioris m.
    - Medial and lateral commissures

- Extrinsic muscles
  - Dorsal rectus m.
  - Ventral rectus m.
  - Medial rectus m.
  - Lateral rectus m.
  - Dorsal oblique m.
    - Trochlea
  - Ventral oblique m.
  - Retractor bulbi m.

- Other external structures
  - Optic n.
  - External ophthalmic a.
  - Lacrimal structures
    - Lacrimal gland
    - Lacrimal lake
    - Lacrimal punctum
  - Zygomatic gland (--CF)
  - Angularis oculi branch of the facial v. (--CF)
  - Periorbita

- Bulbar structures
  - Tunica fibrosa bulbi (fibrous tunic)
    - Cornea
    - Sclera
  - Tunica vasculosa bulbi (vascular tunic)
    - Iris
      - Iridic granules (E---)
      - Pupil
    - Ciliary body
      - Produces aqueous humor
      - Zonular fibers hold the lens
    - Choroid
      - Tapetum lucidum
  - Tunica interna bulbi (internal, nervous tunic)
    - Pars optica retinae
      - Covers most of internal choroid
      - Contains rods and cones
    - Optic disc
    - Ora serrata
    - Pars ceca retinae
      - Pars ciliaris retinae
        - Covers ciliary body
      - Pars iridica retinae
        - Covers caudal part of iris
  - Lens
  - Limbus
  - Iridocorneal angle
    - Area for aqueous humor drainage

Eye (cont.)
- Ocular chambers
  - Camera anterior bulbi (anterior, rostral chamber)
    - Contains aqueous humor
  - Camera posterior bulbi (posterior, caudal chamber)
    - Contains aqueous humor
  - Camera vitrea bulbi (vitreous chamber)
    - Contains the vitreous body and vitreous humor

Ear
- Gross structures (--CF)
  - Helix
  - Tragus
  - Antitragus
- Cartilages
  - Annular
  - Auricular
  - Scutiform
- Auditory canals
  - Vertical
  - Horizontal

CRANIAL NERVES

I. Olfactory n.
- Special sensory
  - Olfaction
- Internal and external foramen
  - Cribiform foramen

II. Optic n.
- Special sensory
  - Vision
- Internal and external foramen
  - Optic canal

III. Oculomotor n.
- Parasympathetic to
  - Sphincter mm. of the iris
  - Ciliary m.
- Motor to
  - Dorsal rectus m.
  - Ventral rectus m.
  - Medial rectus m.
  - Ventral oblique m.
  - Levator palpebrae superioris m.
- Internal and external foramen
  - Orbital fissure (E-CF)
  - Foramen orbitorotundum (-B--)

IV. Trochlear n.
- Motor to
  - Dorsal oblique m.
- Internal and external foramen
  - Orbital fissure (E-CF)
  - Foramen orbitorotundum (-B--)

V. Trigeminal n.
- Ophthalmic n.
  - Sensory to
    - Eye
    - Forehead
    - Frontal sinus
    - Nasal mucosa
  - Internal and external foramen
    - Orbital fissure (E-CF)
    - Foramen orbitorotundum (-B--)
- Maxillary n.
  - Sensory to
    - Nasal mucosa
    - Maxilla
    - Temporal area
    - Horn (-B--)
  - Internal foramen
    - Round foramen
  - External foramen
    - Rostral alar foramen (E-CF)
    - Foramen orbitorotundum (-B--)
- Mandibular n.
  - Sensory to
    - Mandible
    - External ear
  - Motor to
    - Muscles of mastication except caudal digastricus m.
  - Internal and external foramen
    - Oval foramen

IV. Abducens n.
- Motor to
  - Lateral rectus m.
  - Retractor bulbi m.
- Internal and external foramen
  - Orbital fissure

VII. Facial n.
- Special sensory
  - Taste to rostral two-thirds of tongue
    - By way of chorda tympani branch to mandibular n.
- Parasympathetic to
  - Lacrimal gland
  - Mandibular gland
  - Sublingual gland

VII. Facial n. (cont.)
- Motor to
  - Muscles of facial expression
  - Caudal part of digastricus m.
  - Cutaneous colli m. (E---)
- Internal foramen
  - Internal acoustic meatus
- External foramen
  - Stylomastoid foramen

VIII. Vestibulocochlear n.
- Special sensory
  - Vestibular n.
    - Equilibrium
  - Cochlear n.
    - Audition
- Internal foramen
  - Internal acoustic meatus

IX. Glossopharyngeal n.
- Special sensory
  - Taste to caudal one-third of tongue
- Parasympathetic to
  - Parotid gland
  - Zygomatic gland (--CF)
- Sensory to
  - Carotid sinus
  - Middle ear
  - Pharynx
- Motor to
  - Stylopharyngeus m.
- Internal foramen
  - Jugular foramen
- External foramen
  - Occipitotympanic fissure

X. Vagus n.
- Parasympathetic to
  - Abdominal viscera
  - Thoracic viscera
- Sensory to
  - Pharynx
  - Abdominal viscera
  - Thoracic viscera
  - External auditory canal
    - By way of facial n.
- Motor to
  - Larynx
  - Pharynx
- Internal foramen
  - Jugular foramen
- External foramen
  - Occipitotympanic fissure

XI. Accessory n.
- Motor to
  - Trapezius m.
  - Sternocephalicus m.
  - Brachiocephalicus m.
  - Omohyoideus m.
  - Larynx
    - By way of facial n.
- Internal foramen
  - Jugular foramen
- External foramen
  - Occipitotympanic fissure

XII. Hypoglossal n.
- Motor to
  - Tongue
- Internal and external foramen
  - Hypoglossal canal

OTHER STRUCTURES

Salivary glands
- Mandibular gland
  - Location
    - Mandibular angle
  - Duct
    - Course of travel
      - Medial to mandible
    - Opening
      - Sublingual caruncle

- Monostomatic sublingual gland (-BCF)
  - Location
    - Cranial to mandibular gland
  - Opening
    - Sublingual caruncle

- Polystomatic sublingual gland
  - Location
    - Along monostomatic sublingual duct
  - Opening
    - Lateral sublingual recess

- Parotid gland
  - Location
    - Base of the ear
  - Duct
    - Course of travel
      - Ventral to masseter m. (EB--)
      - Lateral to masseter m. (--CF)
    - Opening
      - Next to last premolar (E-CF)

Salivary glands, parotid gland, duct, opening (cont.)
                Next to second molar (-B--)

    Zygomatic gland (--CF)
        Location
            Medial to zygomatic arch
        Openings
            Next to first upper molar

    Buccal glands
        Location
            Medial to buccinator and masseter mm.
        Openings
            Several
        Possibly represent a diffuse zygomatic gland (EB--)

Lymph nodes
    Lateral retropharyngeal ln. (EB--)
        Location
            Near external carotid a.
        Area drained
            Deep structures of head
            Parotid ln.
            Mandibular ln.
            Medial retropharyngeal ln.
        Drains to
            Cranial deep cervical ln. (E---)
            Tracheal ducts (-B--)

    Mandibular ln.
        Location
            Mandibular angle (-BCF)
            Intermandibular space (E---)
        Area drained
            Rostral part of head
            Ventral part of head
            Superficial part of head
        Drains to
            Cranial deep cervical ln. (E---)
            Lateral retropharyngeal ln. (-B--)
            Medial retropharyngeal ln. (E-CF)

    Medial retorpharyngeal ln.
        Location
            Near internal carotid a. (E---)
            Near stylohoid bone (EB--)
            Near diverticulum of auditory tube (E---)
            Medial to mandibular gland (--CF)
        Area drained
            Deep structures of head
            Mandibular ln. (E-CF)
            Parotid ln. (E-CF)

Drains to
- Cranial deep cervical ln. (E---)
- Lateral retropharyngeal ln. (-B--)
- Tracheal duct (-BCF)

Parotid ln.
- Location
  - Cranial to parotid gland (EBc-)
  - Caudal to parotid gland (---f)

Cranial deep cervical ln. (EBc-)
- Location
  - Near thyroid gland
- Area drained
  - Head
  - Neck
  - Mandibular ln. (EB--)
  - Medial retropharyngeal ln. (E-C-)
  - Lateral retropharyngeal ln. (EB--)
- Drains to
  - Tracheal duct
  - Middle deep cervical ln.

Middle deep cervical ln. (EBc-)
- Location
  - Near midcervical region
- Area drained
  - Neck
  - Cranial deep cervical ln.
- Drains to
  - Tracheal duct
  - Caudal deep cervical ln.

Caudal deep cervical ln. (EBcf)
- Location
  - Near rib 1
- Area drained
  - Tracheal duct
  - Superficial cervical ln. (Eb--)
  - Middle deep cervical ln.
  - Axillary ln. (EB--)
  - Scapular region
  - Caudal neck
  - Medial retropharyngeal ln. (---F)
- Drains to
  - Sternal ln. (E---)
  - Tracheal duct (-BCF)
  - Jugular vv. (EBC-)
  - Thoracic duct (EBC-)

Vessels
- Common carotid a.
  - Origin
    - Brachiocephalic a. (--CF)
    - Bicarotid truck (EB--)
  - Cervical location
    - Carotid sheath
      - Internal jugular v. (-BCF)
      - Vagosympathetic n.
  - Gives rise to
    - Internal carotid a. (E-CF)
      - Supplies brain
      - Closely related to diverticulum of auditory tube (E---)
    - External carotid a.
      - Lingual a.
      - Facial a.
      - Caudal auricular a.
      - Maxillary a.

- External jugular v.
  - Boundaries of the cervical jugular furrow
    - Dorsal border
      - Cleidomastoideus m. (EB--)
      - Cleidocervicalis m. (--CF)
    - Ventral border
      - Sternomandibularis m. (EB--)
      - Cleidocervicalis m. (--CF)
    - Medial border
      - Omohoideus m. (E---)
      - Sternomastoideus m. (-B--)
      - Cleidocervicalis m. (--CF)
    - Lateral border
      - Skin
      - Cutaneous colli m. (E---)
  - Formed by
    - Linguofacial v.
      - Anastomosis ventral to larynx with vein of other side (--CF)
      - Lingual v.
        - Sublingual v.
      - Facial v.
        - Drains lateral side of the face
        - Communicates with maxillary v. by way of
          - Orbital branches at angularis oculi (--CF)
          - Buccinator v. (EB--)
          - Deep facial v. (vena reflexa) (EB--)
    - Maxillary v.
      - Joined by facial v.

Horn (-B--)
    Cornual a.
        Origin
            Superficial temporal a.
        Course of travel
            Parallels temporal line
            Dorsal branch
                Runs over base of horn
            Ventral branch
                Runs under base of horn
    Cornual n.
        Origin
            Trigeminal n.
                Ophthalmic n.
                    Zygomaticotemporal n.
        Course of travel
            Immediately ventral to temporal line
                Midway between eye and horn
            May have atypical branches

Nuchal ligament (EBC-) (Fig. 49)
    Funicular nuchae (EBC-)
        Caudal attachment
            Dorsal thoracic spine
            Continuation of supraspinous lig. (E-C-)
        Cranial attachment
            Nuchal crest (EB--)
            Axis (--C-)
        Description
            Paired entire length (--C-)
            Paired but appears single (E---)
            Single caudally, paired cranially (-B--)
        Bursas
            Cranial nuchal bursa (E---)
                Between arch of the atlas and funicular nuchae
            Caudal nuchal bursa (E---)
                Between spine of the axis and funicular nuchae
            Supraspinous bursa (EB--)
                Between spine of T_2 and funicular nuchae
    Lamellar nuchae (EB--)
        Caudal attachment
            Thoracic vertebrae (E---)
        Cranial attachment
            Cervical vertebrae
        Paired

Muscles
    Sternocephalicus m.
        Origin
            Sternum

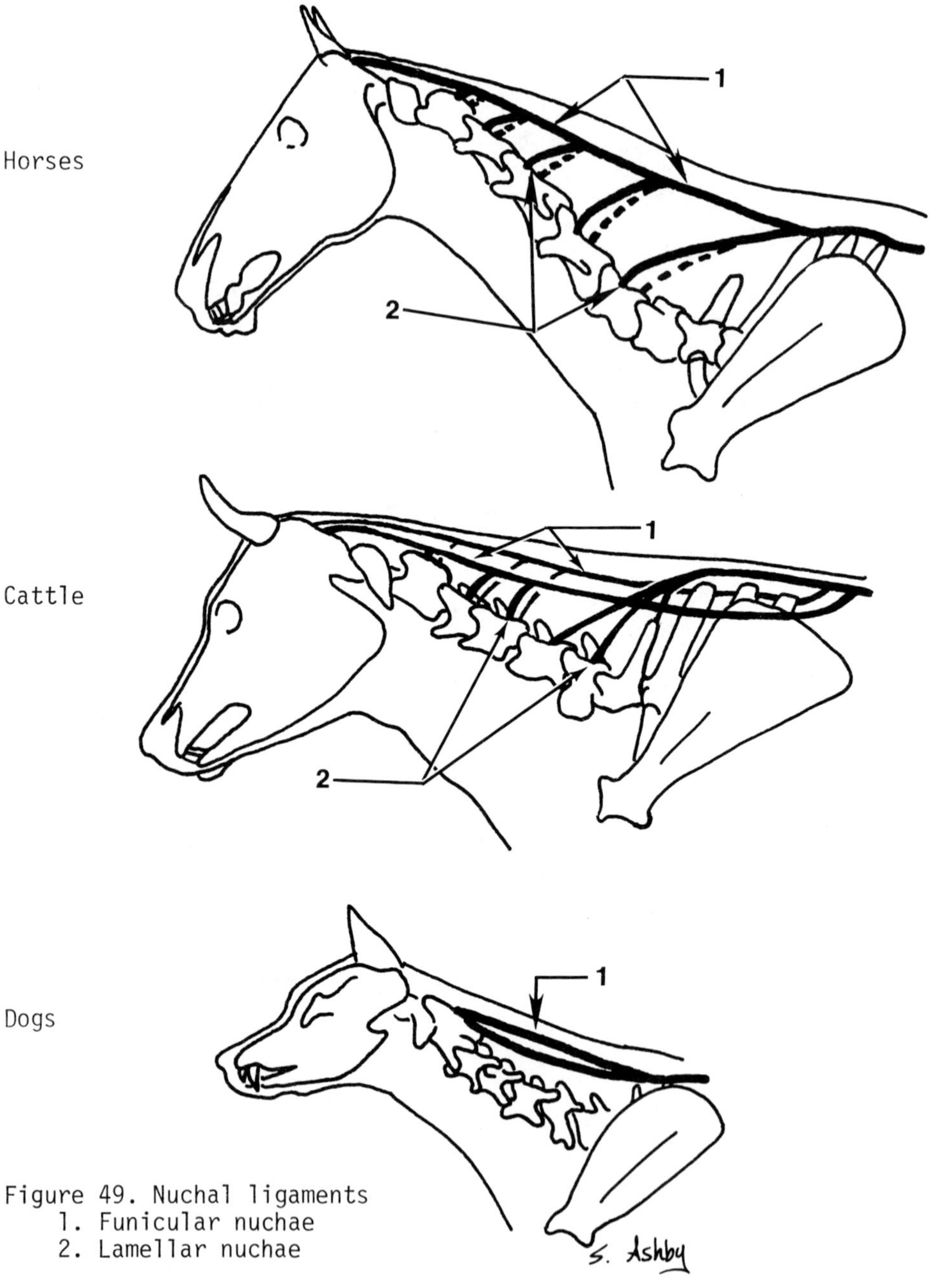

Figure 49. Nuchal ligaments
1. Funicular nuchae
2. Lamellar nuchae

Muscles, sternocephalicus m. (cont.)
- Termination
  - Mandible (sternomandibularis m.) (EB--)
  - Mastoid process (sternomastoideus m.) (-BCF)
  - Occipital bone (sternooccipitalis m.) (--CF)
- Function
  - Lower head and neck
  - Turn head and neck to the side

Sternohyoideus m.
- Origin
  - Sternum
- Termination
  - Basihyoid bone
- Function
  - Pull basihyoid bone and tongue caudally

Sternothyroideus m.
- Origin
  - Sternum
- Termination
  - Thyroid cartilage of larynx
- Function
  - Pull hyoid bones, larynx and tongue caudally

Omohyoideus m. (Eb--)

Esophagus
- Position in midcervical location
  - Dorsolateral to trachea, left side

Thyroid gland
- Paired lobes
- Glandular isthmus (-B--)
- Fibrous isthmus (E--F)

Parathyroid glands
- Caudal
  - Internal
- Cranial
  - External

Crest of the neck

PALPATION

Larynx
- Basihyoid bone
  - Lingual process (E---)
- Thyroid cartilage
- Cricoid cartilage
- Epiglottis (--CF)
- Vocal folds (--CF)
- Cuniform process (--C-)
- Corniculate process (--CF)

- Larynx (cont.)
  - Muscular process (--CF)
  - Swallowing
- Facial crest (E---)
- Infraorbital foramen (E---)
- Mental foramen (E---)
- Facial a. (EB--)
- Palatine tonsil (--CF)
- Salivary duct openings (--CF)
- External jugular v.
  - Linguofacial v.
  - Maxillary v.
  - Normal presystolic pulse (-B--)
- Common carotid a. (E---)
- Trachea
- Sternocephalicus m.
- Sternohyoideus m.
- Lymph nodes
  - Mandibular ln.
  - Parotid ln. (-B--)
  - Medial retropharyngeal ln. (--CF)
- Temporomandibular joint (--CF)
- Mandibular gland (--CF)
- Zygomatic arch
- Occipital protuberance
- Eye
  - Semilunar fold of the conjunctiva (third eyelid)
  - Lacrimal puncta
  - Lacrimal lake
  - Cornea
  - Corneal limbus
  - Iris
    - Iridic granules (E---)
  - Ocular fundus (ophthalmic exam)
  - Palpebrae
    - Cilia
  - (Note the pupil shape)
- Nasal orifice of nasolacrimal duct (E---)
- Nasal diverticulum (E---)
- Ear
  - Auditory canal (--CF)
  - Tympanic membrane (--CF)
- Serratus ventralis m., cervical portion (E---)
- Rhomboideus m., cervical portion (EB--)
- Esophagus
- Atlas
  - Wings

RADIOGRAPHS

- Teeth
  - Alveoli dentales

Nasal conchae (turbinates)
Frontal sinus
Cranial cavity
    Tentorium ossium
Mandible
    Coronoid process
Tympanic bulla
Hard palate
Soft palate
Hyoid bones
    Stylohyoid (Ebcf)
Larynx
    Epiglottis
Nasal cavity
Nasal pharynx
Oral cavity
Oropharynx
Temporomandibular joint
Zygomatic arch
    Zygomatic bone
    Temporal bone
    Zygomatic process of frontal bone
External acoustic meatus
External auditory canal
Foramen magnum (--CF)
Atlantooccipital joint
Atlantoaxial joint
Diverticulum of auditory tube (gutteral pouch) (E---)
Trachea
Esophagus

4

VERTEBRAL COLUMN

OSTEOLOGY

Vertebral formula

	Cervical	Thoracic	Lumbar	Sacral
Horses	7	18	6	5
Cattle	7	13	6	5
Dogs	7	13	7	3
Cats	7	13	7	3
Swine	7	14-15	6-7	4

Location of end of the spinal cord

- L_{6-7} (--C-)
- S_1 (-B--)
- S_2 (E---)
- S_3 (---F)

Vertebral processes
- General features
  - Caudal articular process
  - Cranial articular process
  - Intervertebral foramen
  - Lamina
  - Pedicles
  - Spinous process (spine)
  - Transverse process
  - Vertebral arch
  - Vertebral body
  - Vertebral canal
  - Vertebral foramen
- Cervical vertebrae
  - Atlas
    - Alar foramen (EB--P)
    - Alar notch (--CF-)
    - Caudal articular fovea
    - Cranial articular fovea
    - Lateral foramen
    - Wings
    - Transverse foramen (E-CFP)
    - Ventral tubercle
  - Axis
    - Dens

Vertebral processes, cervical vertebrae, axis (cont.)
- Lateral foramen (EB--P)
- Others
  - Transverse foramen
- Thoracic vertebrae
  - Accessory process
  - Caudal costal fovea
  - Cranial costal fovea
  - Mammillary process
  - Lateral foramen (EB--P)
- Lumbar vertebrae
  - Accessory process
  - Lateral foramen (-B--P)
- Sacrum
  - Apex
    - Caudal articular process
  - Base
    - Cranial articular process
    - Promontory
  - Dorsal sacral foramen
  - Lateral sacral crest
  - Median sacral crest
  - Pelvic sacral foramen
  - Sacral canal
- Caudal vertebrae
  - Hemal processes

Anticlinal vertebrae
- T_{11} (--CF)
- T_{13} (-B--)
- T_{16} (E---)

ASSOCIATED STRUCTURES

Ligaments
- Dorsal longitudinal lig.
  - Along dorsal surface of vertebral bodies
- Intercapital ligg.
  - Join heads of rib bone pairs
    - Absent for rib bone pairs 1, 12, 13 (--CF)
- Interspinous ligg.
  - Join spinous processes of adjacent vertebrae
- Intertransverse ligg.
  - Join transverse processes of adjacent lumbar vertebrae
- Intervertebral discs
  - Annulus fibrosus
    - Joins bodies of adjacent vertebrae
    - Located along vertebral body edge
    - Thinner dorsally

- Encloses nucleus pulposus
  - Nucleus pulposus
- Nuchal lig.
- Supraspinous lig.
  - Continuous with nuchal lig.
  - Joins dorsal parts of spinous processes of adjacent vertebrae
- Ventral longitudinal lig.
  - Joins ventral surface of vertebral bodies
- Yellow ligg.
  - Join arches of adjacent vertebrae

Meninges
- Dura mater
  - Epidural space
    - Contains vertebral venous sinuses
- Arachnoid
  - Subarachnoid space
    - Contains spinal fluid
- Pia mater
  - Contains spinal aa.

Vertebral venous sinsuses

Major epaxial muscle groups
- Erector spinae group
  - Iliocostalis group
    - Most lateral group
    - Regional sections
      - Cervical part
      - Thoracic part
      - Lumbar part
  - Longissimus group
    - Central group
    - Regional sections
      - Capital part
      - Atlantal part
      - Cervical part
      - Thoracic part
      - Lumbar part
  - Transversospinalis group
    - Most of the medial group
    - Regional sections
      - Semispinalis m.
        - Capital part
          - Biventer cervicis m.
          - Complexus m.
        - Cervical part
        - Thoracic part
      - Multifidus mm.
      - Rotator mm.

Major hypaxial muscles
- Lumbar region
  - Iliopsoas m.
    - Psoas major m.
    - Iliacus m.
  - Psoas minor m.
- Cervical region
  - Longus colli m.
  - Longus capitis m.
  - (Longus atlantis m.) (-B--)

PALPATION

Wings of the atlas
Individual rib bones
Thoracic spinous processes
Lumbar vertebrae
- Spinous processes
- Transverse processes

Caudosacral joint
Sternocostal joint

RADIOGRAPHS

Cervical structures
- Atlas
  - Wings
  - Vertebral canal
  - Atlantooccipital joint
- Axis
  - Dens
  - Spinous process
  - Articular processes
  - Atlantoaxial joint
- C_3-C_7
  - (Note the various prominences)
  - Intervertebral spaces
  - Vertebral canal
  - Ventral transverse processes of C_6

Thoracic structures (--CF)
- Vertebrae
  - (Note the various prominences)
  - Intervertebral spaces
  - Vertebral canal
  - Rib articulations

Lumbar structures (--CF)
- Vertebrae
  - (Note the various prominences)
  - Intervertebral spaces

Vertebral Column
        Vertebral canal
Sacral structures (--CF)
    Vertebrae
        (Note the various prominences)
        Sacral canal
        Sacroiliac joint

5

THORAX

THORACIC WALL

Osteology
- Rib bones
  - Number
    - 13 pairs (-BCF)
    - 18 pairs (E---)
  - True (sternal) ribs
    - 8 pairs (EB--)
    - 9 pairs (--CF)
  - False (asternal) ribs
  - Costal cartilage
    - Costochondral junction
    - Costal arch
    - Sternocostal joint
- Sternum
  - Manubrium
    - Cartilage of the manubrium (E---)
  - Xiphoid process
    - Xiphoid cartilage
  - 5 other sternebrae (EB--)
  - 6 other sternebrae (--CF)

Muscles
- External intercostal mm.
  - Craniodorsal fiber direction
- Internal intercostal mm.
  - Cranioventral fiber direction

Intercostal aa.
- Location
  - Caudal side of each rib bone
    - Ventral portion sends a branch to cranial side of the same rib bone
  - Connect the aorta and internal thoracic aa.

Intercostal nn.
- Caudal side of each rib bone
- Ventral branches of thoracic nn.

RESPIRATORY SYSTEM

Pleura
- Parietal pleura
  - Costal pleura
  - Diaphragmatic pleura
    - Lines of pleural reflection
      - Cranioventral
        - Eighth costochondral junction
      - Caudodorsal
        - Last rib bone
  - Mediastinal pleura
    - Thick (-B--)
    - Very thin (E-CF)
- Visceral (pulmonary) pleura
- Pleural cavity

Lungs
- Right lung
  - Cranial lobe
    - Cranial part (-B--)
    - Caudal part (-B--)
    - Trachial bronchus (-B--)
  - Middle lobe (-BCF)
  - Caudal lobe
  - Accessory lobe
- Left lung
  - Cranial lobe
    - Cranial part (-BCF)
    - Caudal part (-BCF)
  - Caudal lobe
- Physical features
  - Groove for esophagus
  - Groove for caudal vena cava
  - Cardiac notch
    - Left (EB--)
    - Right
- Basal border of the lung
  - Cranioventral
    - Costochondral junction of sixth rib bone
  - Caudodorsal
    - Proximal part of next to the last intercostal space

CARDIOVASCULAR SYSTEM

Heart
- External surface
  - Coronary groove
  - Interventricular grooves

- Paraconal
- Subsinuosal

- Atrium
  - Auricles
  - Interatrial septum
    - Fossa ovalis
      - Right atrium
    - Intervenous tubercle
      - Right atrium
  - Pectinate mm.

- Atrioventricular valves
  - Left
    - Parietal cusp
    - Septal cusp
  - Right
    - Angular cusp (EB-F)
    - Parietal cusp
    - Septal cusp

- Ventricle
  - Chordae tendinae
  - Conus arteriosus
    - Right ventricle
  - Interventricular septum
  - Papillary mm.
  - Trabecula septomarginalis
    - Right ventricle
    - Well developed (EB--)

- Semilunar valves
  - 3 cusps
  - Pulmonary valve
    - Right ventricle
    - Pulmonary trunk
  - Aortic valve
    - Left ventricle
    - Aorta

- Ossa cordis (-B--)

- Position of the heart
  - Cranial extent
    - Second intercostal space (EB--)
    - Third rib bone (--C-)
    - Fourth intercostal space (---F)
  - Caudal extent
    - Fifth intercostal space (-B--)
    - Sixth intercostal space (E-C-)
    - Seventh rib bone (---F)

- Great vessels
  - Aorta
    - Ascending aorta
      - Coronary aa.
        - Left coronary a.
          - Paraconal interventricular a. (EBCf)
          - Subsinuosal interventricular a. (-BCf)
          - Circumflex branch (EBCf)
        - Right coronary a.
          - Paraconal interventricular a. (---f)
          - Subsinuosal interventricular a. (E---f)
          - Circumflex branch (EBcf)
    - Aortic arch
      - Brachiocephalic trunk
        - Left subclavian a.
          - Arises fron the aorta (--CF)
          - Internal thoracic a.
          - Superficial cervical a.
            - Deltoid branch
            - Prescapular branch
            - Ascending branch (-BCF)
            - Suprascapular a. (--CF)
          - Vertebral a. (E-CF)
          - Costocervical trunk
            - Dorsal scapular a.
            - Deep cervical a. (-BCF)
            - Supreme intercostal a. (EB-F)
            - Thoracic vertebral a. (--C-)
            - Vertebral a. (-B--)
          - Deep cervical a. (E---)
        - Right subclavian a.
          - Internal thoracic a.
          - Superficial cervical a.
            - Deltoid branch
            - Prescapular branch
            - Ascending branch (-BCF)
            - Suprascapular a. (--CF)
          - Vertebral a. (E-CF)
          - Costocervical trunk
            - Dorsal scapular a.
            - Deep cervical a.
            - Supreme intercostal a. (EB-F)
            - Thoracic vertebral a. (--C-)
            - Vertebral a. (-B--)
        - Bicarotid trunk (EB--)
          - Left common carotid a.
          - Right common carotid a.
        - Left common carotid a. (--CF)
        - Right common carotid a. (--CF)

- Descending aorta
  - Thoracic branches
    - Bronchoesophageal a.
    - Dorsal intercostal aa.
    - Dorsal costoabdominal aa.
    - Cranial phrenic aa. (E---)

- Pulmonary trunk
  - Pulmonary aa.
  - Ligamentum arteriosum

- Azygos v.
  - Left azygos v. (-b--)
  - Right azygos v. (EbCF)

- Caudal vena cava

- Cranial vena cava

- Pulmonary vv.

- Coronary sinus
  - Great cardiac (coronary) v.
    - In the coronary groove
    - In the paraconal interventricular groove
  - Middle cardiac v.
    - In the subsinuosal interventricular groove

Pericardium
- Fibrous pericardium
  - Phrenicopericardial lig. (--CF)
  - Sternopericardial lig. (EB--)
- Serous pericardium
  - Parietal pericardium
  - Visceral pericardium (epicardium)
- Pericardial cavity

Endocardium

Circulation patterns (Fig. 50)

OTHER STRUCTURES

Thoracic nerves
- Phrenic n.
  - Supplies motor fibers to
    - Diaphragm
  - Origin
    - C_5 (-BCF)

ADULT CIRCULATION

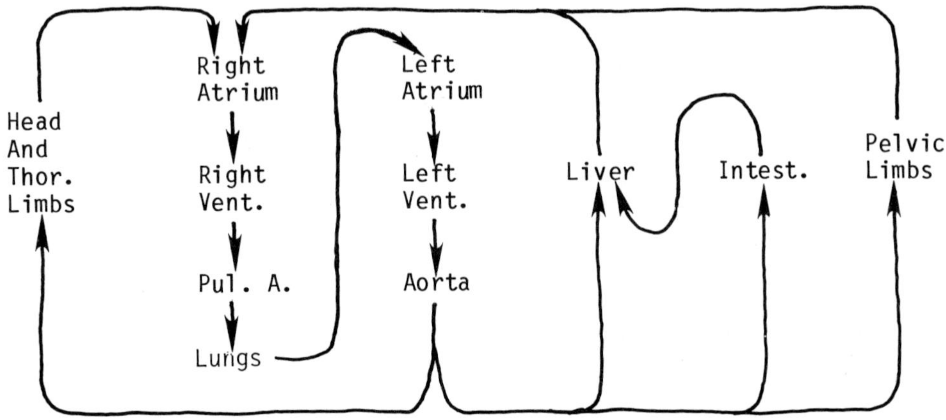

FETAL CIRCULATION

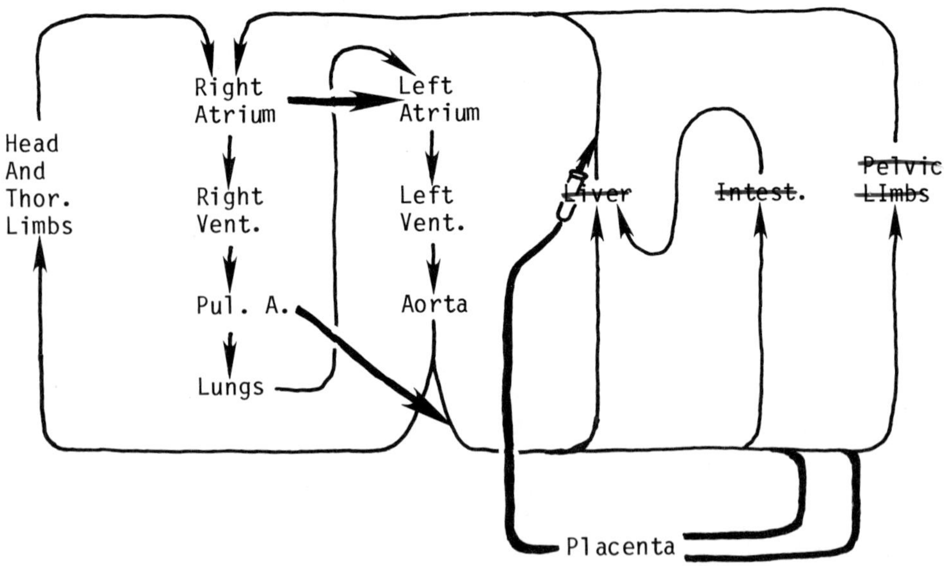

Foramen Ovale----------------------Fossa Ovalis
Ductus Arteriosus------------------Ligamentum Arteriosum
Umbilical Arteries-----------------Round Ligaments of the Urinary Bladder
Ductus Venosus---------------------(No adult remnant)

Figure 50. Circulation Patterns

Thoracic nerves, phrenic n., origin (cont.)
        - C_6
        - C_7
    - Other features
        - Lies in pleural folds
        - Crosses heart about midway ventrally

- Vagus n.
    - Supplies parasympathetic fibers to
        - Abdominal viscera
        - Thoracic viscera
    - Origin
        - Brain
    - Branches
        - Left recurrent laryngeal n.
            - Loops around aortic arch
            - Becomes the left caudal laryngeal n.
        - Right recurrent laryngeal n.
            - Loops around right subclavian a.
            - Becomes the right caudal laryngeal n.
        - Left and right ventral vagal trunks
        - Left and right dorsal vagal trunks
        - Dorsal vagal trunk
        - Ventral vagal trunk
    - Course of travel
        - Crosses base of heart
        - Close to esophagus

Esophagus
- Location
    - Dorsal to trachea
- Passes through esophageal hiatus

Diaphragm
- Parts
    - Muscular portion
    - Central tendon
        - Vena caval foramen
            - Right side of tendon
    - Right crus
        - Esophageal hiatus
            - Esophagus
            - Dorsal vagal trunk
            - Ventral vagal trunk
        - Largest crus
    - Left crus
    - Aortic hiatus
        - Located between right and left crura
        - Aorta
        - Thoracic duct
- Rib attachments are slightly caudal to pleural reflections

Lymphatics
- Thoracic duct
  - Right side caudally (EBC-)
  - Left side caudally (---F)
  - Crosses to left side near T_{5-8} (E-C-)
  - Crosses to left side cranial to heart (-B--)
  - Often paired
  - Empties near termination of left jugular v.
- Tracheobronchial ln.
  - Paired
  - Located at bifurcation of trachea
- Caudal mediastinal ln. (EB--)
  - Between the aorta and esophagus
- Sternal ln.
  - Drains cranial 3 mammae (--CF)
  - May receive lymph from axillary ln.

Thymus
- Location
  - In cranial mediastinum
  - Extends into cervical region (-B--)

PALPATION

Thoracentesis locations
- Immediately dorsal to costochondral junction
- Middle of intercostal space
  - Sixth intercostal space (EB--)
    - Right side only (E---)
  - Seventh intercostal space (EBC-)
  - Eighth intercostal space (--CF)

Auscultation limits of the lungs
- Cranial line
  - Long head of triceps brachii m.
- Dorsal line
  - From caudal border of scapula
  - To wing of ilium
- Caudoventral line
  - From olecranon
  - To last dorsal intercostal space

Auscultation of cardiac valve sounds
- Left atrioventricular valve
  - Left intercostal space
    - 4(-B--)
    - 5(E-CF)
  - Location
    - Caudodorsal to olecranon (E---)
    - Medial to medial epicondyle (-B--)
    - Costochondral junction (--CF)
- Right atrioventricular valve
  - Right intercostal space 3 or 4

- Aortic valve
    - Left intercostal space 4
    - Location
        - Just ventral to a dorsal plane passing through humeral joint
- Pulmonary valve
    - Left intercostal space
        - 3(-B--)
        - 4(E-CF)
    - Location
        - Same level as for left atrioventricular valve

Cardiac puncture or pericardial paracentesis site
- Intercostal space 4 or 5

RADIOGRAPHS

Rib bones
- Costal cartilages

Diaphragm

Heart
- Apex
    - Right ventricle
    - Left ventricle
- Base
    - Right atrium
    - Left atrium
- Sternopericardial lig. (--CF)

Vessels
- Caudal vena cava
- Aorta
- Pulmonary aa.
- Pulmonary vv.

Trachea

Lungs

6

ABDOMEN

ABDOMINAL WALL

External abdominal oblique m.
- Origin
  - Lumbodorsal fascia
  - Caudal rib bones
- Termination
  - Linea alba
  - Cranial pubic lig.
  - Tuber coxae
- Function
  - Abdominal press
- Other features
  - Inguinal lig.
  - Slit in aponeurosis is the superficial inguinal ring
  - Superficial to rectus abdominis m.
  - Craniodorsal fiber direction

Internal abdominal oblique m.
- Origin
  - Tuber coxae
  - Thoracolumbar fascia (-BCF)
- Termination
  - Caudal rib bones
  - Linea alba
  - Cranial pubic lig.
- Function
  - Abdominal press
- Other features
  - Caudal muscle edge forms cranial edge of the deep inguinal ring
  - Cremaster m. is slip of this muscle
  - Superficial to rectus abdominis m.
    - Cranial to the umbilicus where the aponeurosis splits around the rectus abdominis m. (--CF)
  - Cranioventral fiber direction
  - Occupies entire paralumbar fossa (-BCF)

Transversus abdominis m.
- Origin
  - Lumbodorsal fascia
  - Costal arch

Transversus abdominis m. (cont.)
- Termination
  - Linea alba
- Function
  - Abdominal press
- Other features
  - Not far enough caudally to be part of inguinal canal
  - Deep to rectus abdominis m.
    - In caudal 1/3 of the abdomen it is superficial to rectus abdominis m. (--CF)
  - Dorsoventral fiber direction

Rectus abdominis m.
- Origin
  - Costal cartilages
- Termination
  - Pubis
    - By way of cranial pubic lig.
- Function
  - Abdominal press
- Other features
  - Craniocaudal fiber direction

Tunica flava abdominis (EB--)

Inguinal canal
- Superficial inguinal ring
  - Slit in external abdominal oblique m.
- Deep inguinal ring
  - Caudal edge of internal abdominal oblique m.
  - Inguinal lig.
  - Rectus abdominis m.
- Structures passing through the inguinal canal
  - External pudendal a.
  - External pudendal v. (-BCF)
  - Genitofemoral n.
    - Spinal nerve contributions
      - L_2 (eb--)
      - L_3 (EBC-)
      - L_4 (-bCF)
  - Vaginal process
    - Vaginal ring
    - Round lig. of the uterus (female) (--CF)
    - Vaginal tunics (male)
      - Enclose structures of spermatic cord
  - Cremaster m. (male)
    - Passes through superficial inguinal ring only
  - Lymphatic vessels

Femoral canal
- Structures located in it
    - Femoral a.
    - Femoral v.
    - Saphenous n.
    - Iliofemoral (deep inguinal) ln. (E---)

DIGESTIVE SYSTEM

Stomach
- Simple stomach (E-CFP)
    - Cardia
        - Saccus cecus (blind sac) (E----)
    - Fundus
        - Diverticulum (----P)
    - Body
        - Margo plicatus (E----)
    - Pylorus
        - Pyloric sphincter
        - Torus pyloricus (-B--P)
    - Approximate volume
        - 1-10 liters (--CFP)
            - 2.5 liters average (--C--)
        - 10-15 liters (E----)
- Ruminant stomach (-B---)
    - Rumen
        - Ruminoreticular groove and fold
        - Cranial and caudal grooves and pillars
        - Left and right longitudinal grooves and pillars
            - Left and right accessory grooves and pillars
            - Insula ruminis
        - Dorsal and ventral coronary grooves and pillars
        - Cranial sac (atrium)
        - Dorsal sac
        - Ventral sac
        - Caudodorsal blind sac
        - Caudoventral blind sac
        - Cardia
        - Ruminoreticular ostium
        - Approximate volume
            - 100-150 liters
    - Reticulum
        - Ruminoreticular groove and fold
        - Reticular groove
        - Reticuloomasal ostium
        - Approximate volume
            - 5 liters

Stomach, ruminant stomach (cont.)

- Omasum
    - Omasal groove
    - Approximate volume
        - 10 liters
- Abomasum
    - Omasoabomasal orifice
    - Approximate volume
        - 15 liters

Small intestine

- Duodenum
    - Cranial duodenum
    - Sigmoid loop (EB--P)
        - Opening of the bile duct (-B---)
    - Ampulla duodeni (E----)
    - Cranial duodenal flexure
    - Descending duodenum
        - Opening of the bile duct (E-CFP)
        - Opening of pancreatic duct(s)
    - Caudal duodenal flexure
    - Ascending duodenum
        - Passes from right to left caudal to root of the mesentery
    - Duodenojejunal flexure
    - Approximate length
        - 100 cm (EB---)
        - 80 cm (----P)
        - 50 cm (--C--)
        - 10 cm (---F-)
- Jejunum
    - Approximate length
        - 40 m (-B---)
        - 35 m (E----)
        - 18 m (----P)
        - 3 m (--C--)
        - 1 m (---F-)
- Ileum
    - Approximate length
        - 70 cm (EB---)
        - 50 cm (----P)
        - 20 cm (--C--)
        - 15 cm (---F-)

Large intestine (Figs. 51,52)

- Junction of large and small intestines
    - Ileocecal orifice (E----)
    - Ileocecocolic orifice (-B--P)
    - Ileocolic orifice (--CF-)
    - Cecocolic orifice (E-CF-)

Cecum
- Located on right side
- Tenia (bands) (E---P)
  - 3 (----P)
  - 4(E----)
- Haustra (E---P)
- Approximate length
  - 100 cm (E----)
  - 75 cm (-B---)
  - 25 cm (----P)
  - 20 cm (--C--)
  - 5 cm (---F-)
- Approximate volume
  - 30 liters (E----)
  - 8 liters (-B---)

Colon
- Ascending colon
  - Right ventral colon (E----)
    - 4 tenia (bands)
    - Haustra
  - Sternal flexure (E----)
  - Left ventral colon (E----)
    - 4 tenia (bands)
    - Haustra
  - Pelvic flexure (E----)
    - 1 tenia (band)
  - Left dorsal colon (E----)
    - 1 to 3 tenia (bands)
  - Diaphragmatic flexure (E----)
  - Right dorsal colon (E----)
    - 3 tenia (bands)
  - Proximal loop (-B---)
  - Spiral loop (ansa spiralis) (-B--P)
    - Centripetal turns
    - Central flexure
    - Centrifugal turns
  - Distal loop (-B---)
  - Approximate length
    - 900 cm (-B---)
    - 400 cm (E---P)
    - 10 cm (--C--)
    - 6 cm (---F-)
  - Approximate volume
    - 60 to 120 liters (E----)

- Transverse colon
  - 2 tenia (bands) (E----)
  - Haustra (E----)

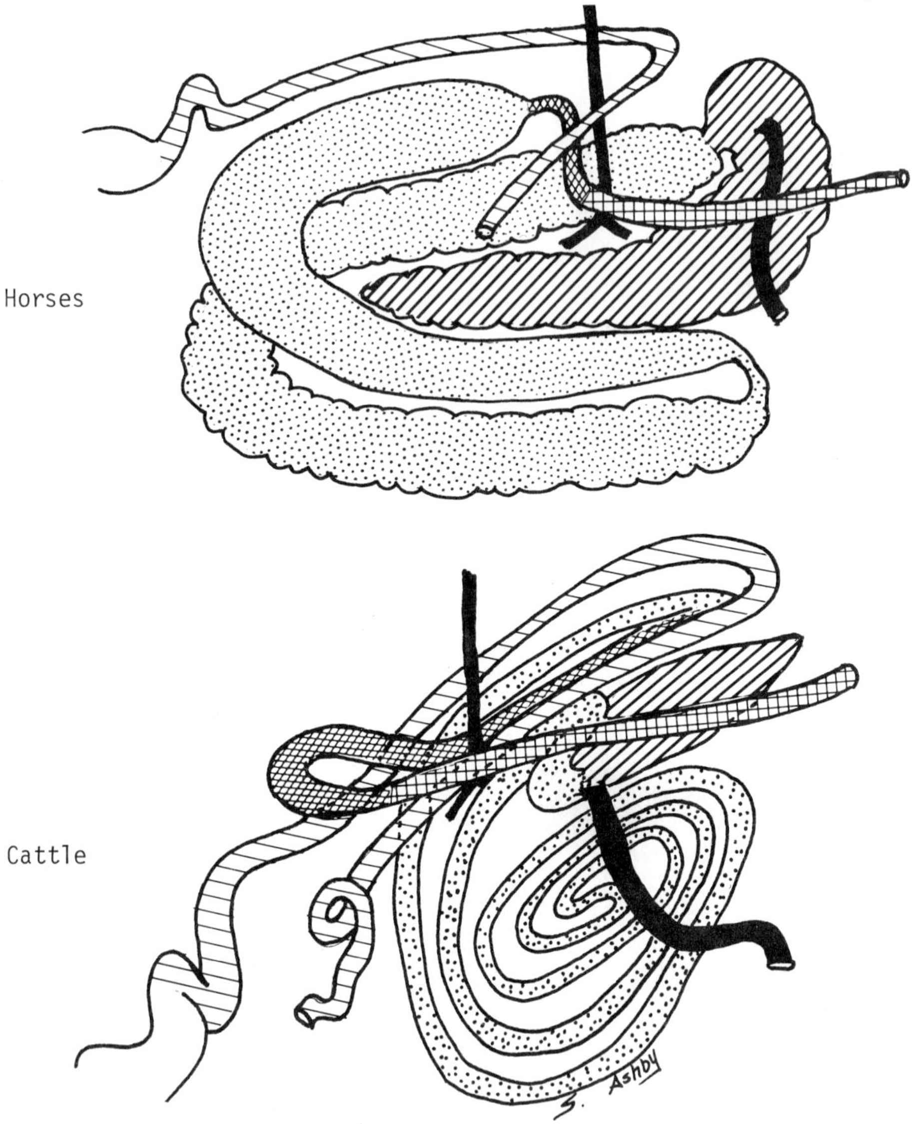

Figure 51. Equine and bovine digestive systems (left, craniodorsal aspect)

Dogs and cats

Pigs

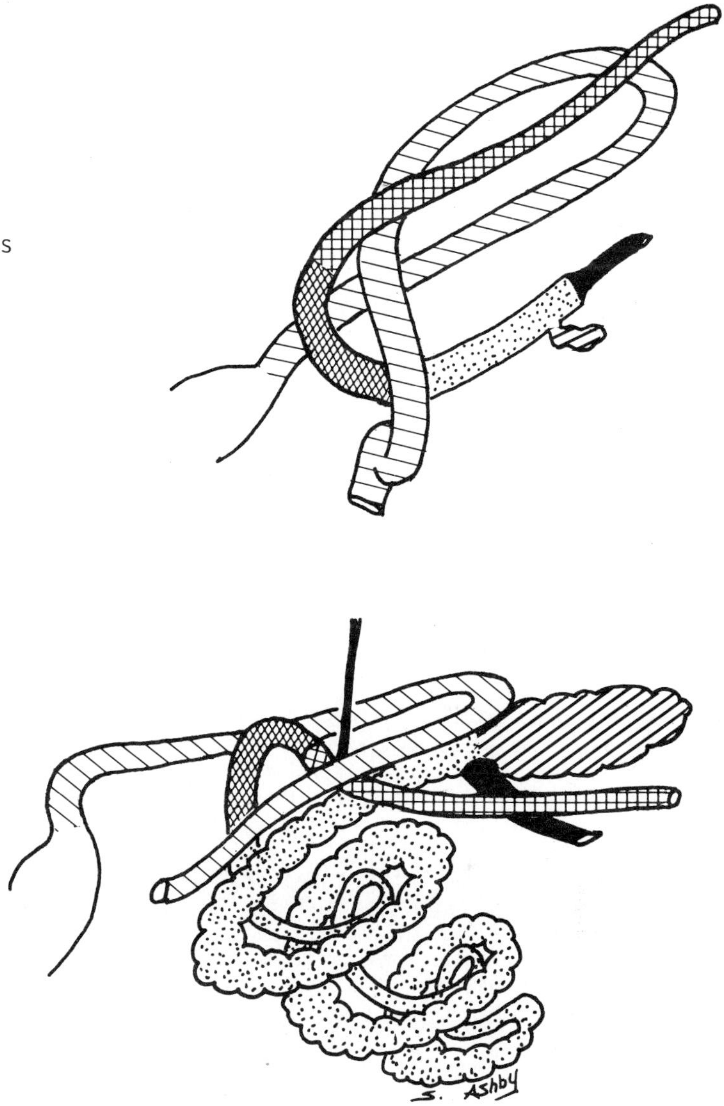

Figure 52. Canine, feline and porcine digestive systems (left, craniodorsal aspect)

- Duodenum
- Ileum
- Cecum
- Ascending colon
- Transverse colon
- Descending colon

Large intestine, colon, transverse colon (cont.)
- Approximate length
  - 40 cm (EB---)
  - 30 cm (----P)
  - 20 cm (--C--)
  - 8 cm (---F-)

- Descending colon
  - 2 tenia (bands) (E----)
  - Haustra (E----)
  - Approximate length
    - 350 cm (E----)
    - 100 cm (-B---)
    - 60 cm (----P)
    - 30 cm (--C--)
    - 20 cm (---F-)

- Rectum

- Anus
  - Anal canal
  - External anal sphincter
  - Internal anal sphincter

Liver
- Lobes
  - Left hepatic lobe
    - Left medial hepatic lobe (E-CFP)
    - Left lateral hepatic lobe (E-CFP)
    - Everything left of a line from round lig. to esophagus
  - Right hepatic lobe
    - Right medial hepatic lobe (--CFP)
    - Left lateral hepatic lobe (--CFP)
    - Everything right of a line from gall bladder location to caudal vena cava
  - Quadrate lobe
    - Between left and right hepatic lobes
    - Ventral to hepatic porta
  - Caudate lobe
    - Caudate process
    - Papillary process (-BCF-)
    - Between left and right hepatic lobes
    - Dorsal to hepatic porta

- Gall bladder (-BCFP)

- Biliary tree
  - Bile duct
    - Opens in sigmoid loop (-B---)
    - Opens in descending duodenum (E-CFP)
    - Opens with the pancreatic duct (E-CF-)

- Cystic duct
- Hepatic ducts

Location
- Cranial abdominal cavity
  - Right side of abdomen only (-B---)

Pancreas
- In major (greater) omentum
- Ducts
  - Pancreatic duct (E-CF-)
    - Opens with bile duct on the major duodenal papilla
  - Accessory pancreatic duct
    - Opens distal to the bile duct on the minor duodenal papilla

ABDOMINAL VESSELS

Arteries
- Phrenicoabdominal trunk (--CF)
  - Caudal phrenic a.
  - Cranial abdominal a.

- Celiac a. (Fig. 53)
  - Left gastric a.
    - Supplies
      - Minor (lesser) curvature of stomach (E-CF)
      - Minor (lesser) curvature of abomasum (-B--)
    - Left gastroepiploic a. (-B--)
      - Supplies
        - Major (greater) curvature of abomasum
  - Hepatic a.
    - Supplies
      - Liver
    - Right gastric a.
      - Supplies
        - Minor (lesser) curvature of stomach (E-CF)
        - Minor (lesser) curvature of abomasum (-B--)
    - Gastroduodenal a.
      - Right gastroepiploic a.
        - Supplies
          - Major (greater) curvature of stomach (E-CF)
          - Major (greater) curvature of abomasum (-B--)
      - Cranial pancreaticoduodenal a.
  - Splenic a.
    - Supplies
      - Spleen
    - Left gastroepiploic a. (E-CF)
      - Supplies
        - Major (greater) curvature of stomach

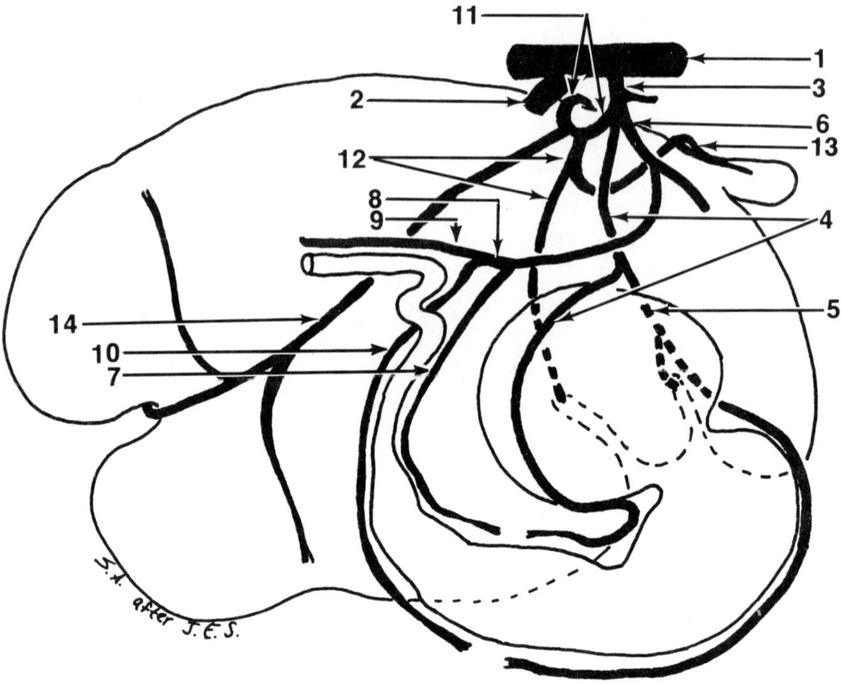

Figure 53. Arteries of the rumen (right aspect)

1. Aorta
2. Cranial mesenteric a.
3. Celiac a.
4. Left gastric a.
5. Left gastroepiploic a.
6. Hepatic a.
7. Right gastric a.
8. Gastroduodenal a.
9. Cranial pancreatoduodenal a.
10. Right gastroepiploic a.
11. Splenic a.
12. Left ruminal a.
13. Reticular a.
14. Right ruminal a.

Arteries, celiac a., splenic a. (cont.)

- Short gastric aa. (E-CF)
  - Supplies
    - Major (greater) curvature of stomach
- Right ruminal a. (-B--)
- Left ruminal a. (-B--)
  - Reticular a.

Caudal phrenic aa. (-B--)

Cranial mesenteric a.
- Supplies
  - Small intestines
- Caudal pancreaticoduodenal a.
- Middle colic a.
  - Supplies
    - Transverse colon
    - Descending colon (E-CF)
    - Ascending colon (-B--)
- Right colic a.
  - Supplies
    - Ascending colon
    - Transverse colon (--CF)
- Ileocolic a.
  - Colic branch
  - Cecal a.
    - Ileal branch

Renal aa.

Testicular aa. (males)

Ovarian aa. (females)

Caudal mesenteric a.
- Left colic a.
  - Supplies
    - Descending colon
- Cranial rectal a.
  - Supplies
    - Rectum
    - Descending colon

Lumbar aa.

Deep circumflex iliac aa. (--CF)

External iliac aa. (Figs. 54,55)
- Supplies
  - Medial and distal pelvic limb
  - Caudal abdominal muscles
  - Ventral genitalia

Arteries (cont.)
- Internal iliac aa. (Figs. 54,55)
  - Supplies
    - Proximolateral pelvic limb
    - Pelvic organs
    - Caudal genitalia
- Median sacral a.

Veins
- Portal v.
  - Gastroduodenal v.
  - Splenic v.
  - Cranial mesenteric v.
  - Caudal mesenteric v.
- Caudal vena cava
  - Common iliac vv.
  - Lumbar vv.
  - Deep circumflex iliac vv. (--CF)
  - Testicular v. (males)
    - Right vein enters into or near right renal v. (--CF)
  - Ovarian v. (females)
    - Right vein enters into or near right renal v. (--CF)
  - Renal vv.
  - Suprarenal vv. (EB--)
  - Caudal phrenic vv. (EB--)
  - Cranial abdominal vv. (EB--)
  - Phrenicoabdominal vv. (--CF)
    - Caudal phrenic v.
    - Cranial abdominal v.
  - Hepatic vv.

OTHER STRUCTURES

Peritoneum
- Parietal peritoneum
- Visceral peritoneum
- (Connecting peritoneum)
  - Mesentery
  - Major (greater) omentum
    - Attaches to
      - Major (greater) curvature of stomach (E-CF)
      - Longitudinal grooves (-B--)
      - Duodenum (-B--)
      - Pancreas
      - Spleen (E-CF)
        - Gastrosplenic lig.
    - Omental bursa
      - Epiploic foramen

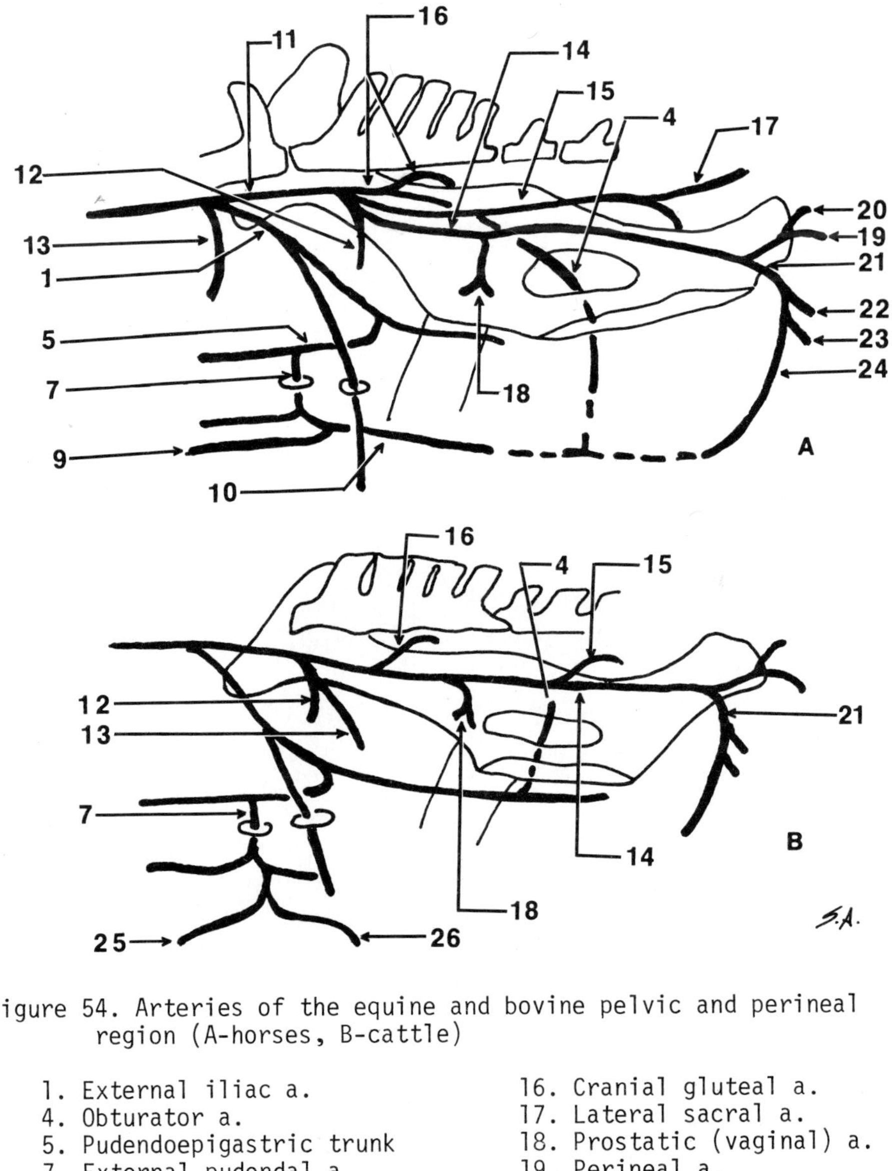

Figure 54. Arteries of the equine and bovine pelvic and perineal region (A-horses, B-cattle)

1. External iliac a.
4. Obturator a.
5. Pudendoepigastric trunk
7. External pudendal a.
9. Cranial a. of penis
10. Middle a. of penis
11. Internal iliac a.
12. Umbilical a.
13. (Uterine a.)
14. Internal pudendal a.
15. Caudal gluteal a.
16. Cranial gluteal a.
17. Lateral sacral a.
18. Prostatic (vaginal) a.
19. Perineal a.
20. Caudal rectal a.
21. Artery of penis
22. Artery of bulb
23. Deep a. of penis
24. Dorsal a. of penis
25. (Cranial mammary a.)
26. (Caudal mammary a.)

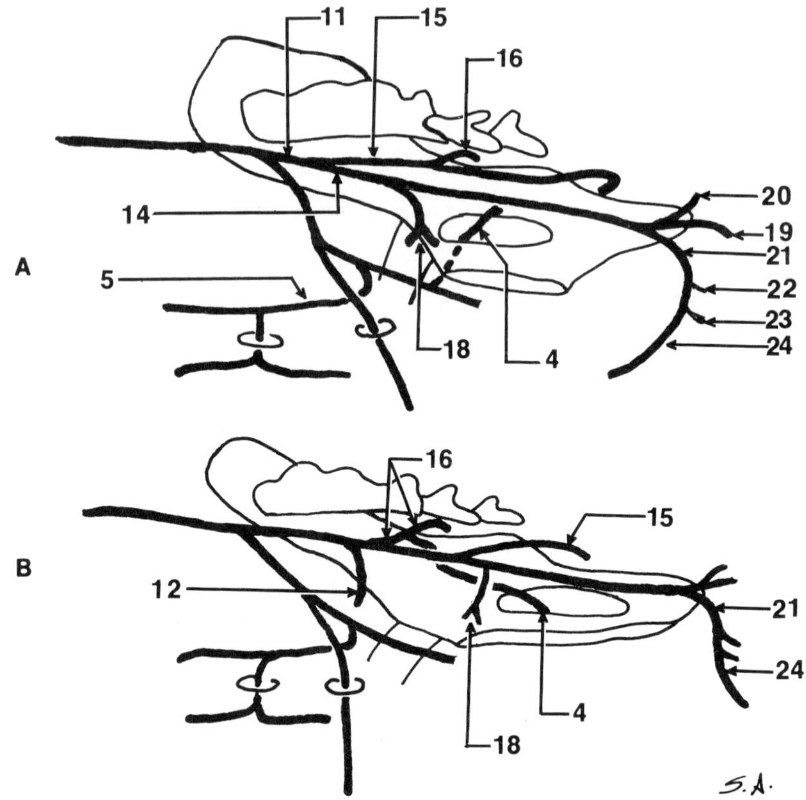

Figure 55. Arteries of the canine and feline pelvic and perineal region (A-dogs, B-cats)

4. Obturator a.
5. Pudendoepigastric trunk
11. Internal iliac a.
12. Umbilical a.
14. Internal pudendal a.
15. Caudal gluteal a.
16. Cranial gluteal a.
18. Prostatic (vaginal) a.
19. Perineal a.
20. Caudal rectal a.
21. Artery of penis
22. Artery of bulb
23. Deep a. of penis
24. Dorsal a. of penis

Peritoneum, (connecting peritoneum) (cont.)
- Minor (lesser) omentum
  - Attaches to
    - Minor (lesser) curvature of stomach (E-CF)
    - Omasum (-B--)
    - Abomasum (-B--)
    - Liver
      - Hepatogastric lig.
      - Hepatoduodenal lig.
- Peritoneal cavity
- Caudal peritoneal reflections
  - Rectogenital pouch
    - Pararectal fossae
  - Vesicogenital pouch
  - Pubovesical pouch
- Peritoneal ligaments
  - Falciform lig.
  - Coronary lig. of the liver
  - Right and left triangular ligg. of the liver
  - Hepatorenal lig.
  - Duodenocolic fold
  - Ileocecal fold
  - Cecocolic fold (E---)
  - Genital fold (males)

Suprarenal (adrenal) glands
- Located craniomedial to kidney

Spleen
- Located
  - Left side
  - Caudal to stomach (E-CF)
  - Cranial to rumen (-B--)
- Hilus

Lymph nodes
- Jejunal ln.
  - Location
    - Mesentery
  - Area drained
    - Intestines
  - Drains to
    - Cisterna chyli

- Medial iliac ln.
  - Location
    - Near origin of external iliac a.
    - Near origin of internal iliac a. (-B--)
    - Near origin of deep circumflex iliac a. (E-C-)

Lymph nodes, medial iliac ln. (cont.)
        Area drained
            Lumbosacral area
            Urogenital area
            Pelvic limb
            Subiliac ln. (EB-F)
            Lateral iliac ln. (EB--)
        Drains to
            Cisterna chyli

    Lateral iliac ln. (EB--)
        Location
            Near deep circumflex iliac a.
        Area drained
            Abdomen
            Subiliac ln.
        Drains to
            Medial iliac ln.
            Lumbar ln.

    Iliofemoral (deep inguinal) ln.
        Location
            Near external iliac a.
            Near origin of deep circumflex iliac a.
            Near deep inguinal ring (E---)
        Area drained
            Inguinofemoral ln.
            Popliteal ln.
        Drains to
            Cisterna chyli (-B--)
            Medial iliac ln.

    Inguinofemoral (superficial inguinal) ln.
        Location
            Near origin of caudal superficial epigastric a.
        Area drained
            Inguinal area
        Drains to
            Iliofemoral ln. (E---)
            Medial iliac ln. (-BCF)
            Lateral iliac ln. (B--)

UROGENITAL SYSTEM

Urinary system
    Kidney
        Location
            Right kidney
                T_{12}-L_2 (--C--)

T_{13}-L_3 (-B---)
T_{15}-T_{17} (E----)
L_1-L_4 (----P)
L_2-L_3 (---F-)
Left kidney
T_{13}-L_3 (--C--)
T_{17}-L_3 (E----)
L_1-L_4 (----P)
L_2-L_3 (-B-F-)
On the midline (-B---)
Structures
Cortex
Medulla
Hilus
Renal sinus (E-CFP)
Renal pelvis
Terminal recesses (E----)
Renal calyces (-B---)
Renal pyramids
Renal crest (E-CF-)

Ureters

Urinary bladder
Ureteral openings
Trigon
Peritoneal ligaments
Median lig.
May contain vestige of urachus
Lateral ligg.
Round ligg.
Umbilical aa. vestiges

Female genital system
Ovary
Size
8 x 4 x 4 cm (E----)
5 x 3 x 3 cm (----P)
3 x 2 x 1.5 cm (-B---)
2 x .5 x .5 cm (--C--)
1 x .5 x .5 cm (---F-)
Location
Right overy is more cranial (--CF-)
Caudolateral to kidney (--CF-)
Ovarian fossa (E----)

Female genital system (cont.)
- Uterine tubes
  - Infundibulum
    - Fimbriae
  - Abdominal ostium
  - Uterine ostium

- Uterus
  - Right and left uterine horns
    - Fetus usually carried here (-BCFP)
  - Uterine body
    - Fetus usually carried here (E----)
  - Uterine cervix
    - External uterine os
    - Internal uterine os
    - Transverse folds (-B---)
    - Longitudinal folds (EBcf-)
    - Pulvini cervicales (cervical prominances) (----P)

- Vagina
  - Fornix

- Vestibule
  - Urethral opening
  - Suburethral diverticulum (-B--p)
  - Vestibular bulbs (E-C--)
  - Minor vestibular gland openings (E-C-P)
  - Major vestibular gland openings (-B-F-)
    - Openings are near urethral opening
  - Urethral tubercle (--C--)

- Vulva
  - Clitoris
    - Body
    - Glans clitoridis (EBCfp)
    - Clitorial fossa (EBCfP)
  - Labia
    - Dorsal commissure
    - Ventral commissure

- Ligaments
  - Suspensory lig. (--CF-)
    - Attaches ovary to body wall
    - Caudolateral to kidney
  - Proper lig. of the ovary
    - Attaches ovary to uterus
  - Intercornual lig. (-B---)
    - Connects uterine horns
  - Broad lig.
    - (Connecting peritoneum)

- Mesovarium
    - Attaches ovary to body wall
    - Contains ovarian vessels
    - Suspensory lig. is in the edge (--CF-)
- Mesosalpinx
    - Attaches uterine tube to mesovarium
        - Forms ovarian bursa
        - Entrance to ovarian bursa
- Mesometrium
    - Attaches uterus to body wall
    - Round ligg. of the uterus
        - Extends from tip of uterine horn to deep inguinal ring
            - Extends into vaginal process (--CF-)

Mammary gland

- Structure
    - Lactiferus duct
    - Lactiferus sinus
        - Glandular part
        - Papillary part
    - Papillary duct (teat canal)
    - Papillary opening

- Equine mammary glands
    - 2 glands
        - 2 ducts for each papilla
    - Blood supply
        - External pudendal a. & v.
    - Lymphatics
        - Inguinofemoral (superficial inguinal) ln.

- Bovine mammary glands
    - 4 glands
    - Suspensory apparatus (Fig. 56)
        - Lateral lamina of the udder
            - Origin is symphyseal (subpelvic) tendon
            - Paired
            - Divides into several leaves
            - Cranioventral fiber direction
            - Made of fibrous connective tissue
        - Medial lamina of the udder
            - Origin is tunica flava abdominis
                - Internal abdominal wall
                - Transverse abdominal fascia
            - Paired
            - Continuous with cranial pubic lig.
            - Made of elastic connective tissue
                - Empty udder allows teats to hang vertically
                - Full udder stretches medial lig. so teats point laterally

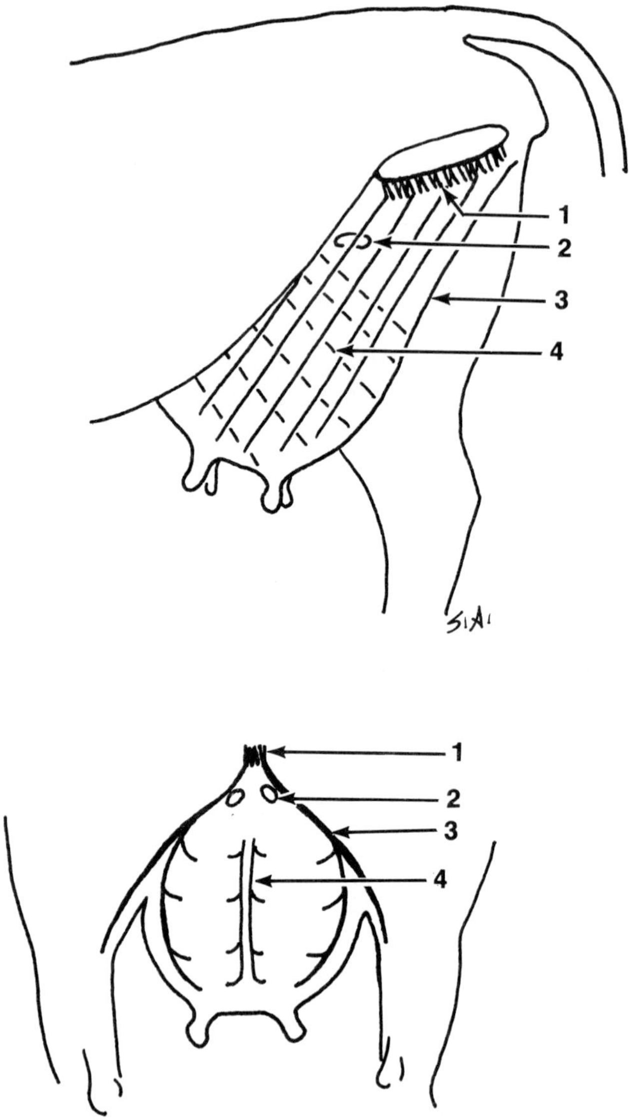

Figure 56. Ligaments of the bovine udder (A-lateral aspect, B-caudal aspect)

1. Symphyseal tendon
2. Inguinofemoral ln.
3. Lateral lamina
4. Medial lamina

Mammary gland, bovine mammary glands, suspensory apparatus, medial lamina of the udder (cont.)

- Craniodorsal fiber direction
- Joins to leaves of the lateral lamina
- Blood supply (Figs. 54,57)
  - Arteries
    - Perineal a.
    - External pudendal a.
  - Veins
    - Perineal v.
      - Valves normally allow blood flow toward gland
    - External pudendal v.
    - Caudal superficial epigastric v.
      - Returns to cranial superficial epigastric v. which returns to internal thoracic v.
    - Form venous circle
- Lymphatics
  - Inguinofemoral (superficial inguinal) ln.

Canine and feline mammary glands

- Usually 5 pair of glands
- Blood supply
  - Cranial superficial epigastric a. & v.
  - Caudal superficial epigastric a. & v.
- Lymphatics
  - Glands 1, 2, and sometimes 3
    - Drain to
      - Axillary ln.
      - Sternal ln.
  - Glands 4,5, and sometimes 3
    - Drain to
      - Inguinofemoral (superficial inguinal) ln.

Porcine mammary glands

- Usually 5 to 7 pair of glands
  - 2 ducts for each papilla

Male genital system

- Testis
  - Size
    - 10 x 7 x 7 cm (-B---)
    - 10 x 6 x 5 cm (E---P)
    - 3 x 2 x 2 cm (--C--)
    - 2 x 1.5 x 1.5 cm (---F-)
  - Structure
    - Tunica albuginea
    - Head end
    - Tail end
  - Age of descent through the inguinal canal
    - 100 days of gestation (-B---)

Male genital system (cont.)
- Urethra and penis
  - Pelvic urethra
    - Colliculus seminalis
      - Ejaculatory orifice
    - Uterus masculinus (Ebcfp)
  - Spongy urethra

  - Penis
    - Root of the penis
      - Retractor penis m.
        - Paired
        - Passes through bulbospongiosus m. (E----)
        - Extends from anus to glans penis (EBCF-)
        - Extends from anus to distal part of sigmoid flexure (----P)
      - Bulbospongiosus m.
        - Unpaired
        - Continues to the glans penis (E----)
        - Bulbus penis (erectile tissue)
          - Origin is under the muscle
          - Continues as the corpus spongiosum penis
            - Surrounds urethra
            - Continues as the corpus spongiosum glandis (E-CF-)
      - Ischiocavernosus m.
        - Paired
        - Extends from ischial arch to penis
        - Right and left crura (erectile tissue)
          - Origin is under the muscle
          - Becomes the corpus cavernosum penis
            - Dorsal to urethra
            - Distinctly paired (--CF-)
      - Suspensory ligg. of the penis

    - Body of the penis
      - Sigmoid flexure (-B--P)

    - Free part of the penis
      - Glans penis
        - Collum glandis (E----)
        - Corona glandis (E----)
        - Fossa glandis (E----)
          - Urethral sinus
        - Raphe (-B---)
        - Pars longa glandis (--C--)
        - Bulbus glandis (--C--)
        - Urethral process
        - Os penis (--CF-)
          - Partially surrounds urethra (--C--)
        - Penile spines (---F-)
          - Project cranially

Prepuce (Fig. 58)
- External layer
    - Continuous caudally with skin
- Preputial ostium (orifice)
- Internal layer
- Preputial fold (E----)
    - External layer
    - Internal layer
    - Preputial ring
- Preputial diverticulum (----P)
- Cranial preputial m. (-BCFP)
- Caudal preputial m. (-Bcfp)

Homologues

Male structures	Female structures
Testis	Ovary
Testicular bursa	Ovarian bursa
Gubernaculum testis	Round lig. of the uterus Proper lig. of ovary
Proper lig. of the testis	Proper lig. of ovary
Retractor penis m.	Retractor clitoridis m.
Bulbospongiosus m.	Constrictor vestibuli m. Constrictor vulvae m.
Bulbus penis	Vestibular bulbs
Ischiocavernosus m.	Ischiocavernosus m.
Penis	Clitoris
Crura of the penis	Crura of the clitoris
Corpus cavernosum penis	Body of the clitoris
Glans penis	Glans clitoridis
Os penis	Os clitoridis
Prepuce	Prepuce of the clitoris (fold of the fossa clitoridis)
Scrotum	Labia

PALPATION

Auscultate normal intestinal sounds
- Cecum (EB--)
- Colon
- Small intestine
- Rumen (-B--)

Spleen (--CF)

Kidneys

Rumen (-B--)

Intestines
- Cecum (E---)
- Descending colon

Urinary bladder

Pudendal n. (-B--)
- Medial to broad sacrotuberous lig.

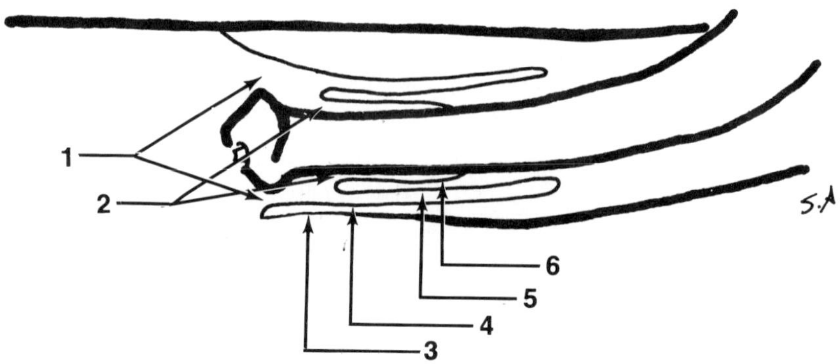

Figure 58. Layers of the equine prepuce

1. Preputial ostium
2. Preputial ring
3. External layer of prepuce
4. Internal layer of prepuce
5. External layer of preputial fold
6. Internal layer of preputial fold

Pudendal n. (cont.)
    Distal cutaneous branch
        Medial to ischiatic tuberosity
Female structures
    Ovaries (EB--)
    Uterus
        Body
        Horns
    Vagina
    Vestibule
        Suburethral diverticulum (-B--)
    Vulva
Male structures
    Testes
    Penis
    Prepuce
    Pelvic urethra (EBC-)
    Prostate (EBC-)
    Seminal vesicles (E---)
    Vesicular gland (-B--)
    Ampulla of the ductus deferens (EB--)
    Bulbourethral glands (EB--)
    Vaginal and deep inguinal rings (E---)

RADIOGRAPHS

Digestive organs
    Stomach
    Small intestines
    Colon
    Liver
Spleen
Urogenital system
    Kidneys
    Urinary bladder
    Prostate
    Mammary papillae
    Prepuce
    Os penis
        Cat: calcified only in older cats
    Scrotum

INDEX